Concurrent Engineering and Design for Manufacture of Electronics Products

Concurrent Engineering and Design for Manufacture of Electronics Products

SAMMY G. SHINA
University of Lowell

VNR VAN NOSTRAND REINHOLD
New York

Copyright © 1991 by Van Nostrand Reinhold

Library of Congress Catalog Card Number 90-23616
ISBN 0-442-00616-0

All rights reserved. No part of this work covered by
the copyright hereon may be reproduced or used in any
form by any means—graphic, electronic, or
mechanical, including photocopying, recording, taping,
or information storage and retrieval systems—without
written permission of the publisher.

Printed in the United States of America.

Van Nostrand Reinhold
115 Fifth Avenue
New York, New York 10003

Chapman and Hall
2-6 Boundary Row
London, SE1 8HN, England

Thomas Nelson Australia
102 Dodds Street
South Melbourne 3205
Victoria, Australia

Nelson Canada
1120 Birchmount Road
Scarborough, Ontario MIK 5G4, Canada

16 15 14 13 12 11 10 9 8 7 6 5 4 3 2

Library of Congress Cataloging-in-Publication Data

Shina, Sammy G.
    Concurrent engineering and design for manufacture of electronics
  products / Sammy G. Shina.
      p.   cm.
    Includes index.
    ISBN 0-442-00616-0
    1. Electronic industries. 2. Electronic apparatus and appliances—
  —Design and construction. 3. Production engineering.   I. Title.
TK7836.S48   1991
621.381—dc20   90-23616
                                                                        CIP

**TO
JACKIE, MIKE, GAIL, NANCY, AND JON**

Preface

This book is intended to introduce and familiarize design, production, quality, and process engineers, and their managers to the importance and recent developments in concurrent engineering (CE) and design for manufacturing (DFM) of new products.

CE and DFM are becoming an important element of global competitiveness in terms of achieving high-quality and low-cost products. The new product design and development life cycle has become the focus of many manufacturing companies as a road map to shortening new product introduction cycles, and to achieving a quick ramp-up of production volumes. Customer expectations have increased in demanding high-quality, functional, and user-friendly products. There is little time to waste in solving manufacturing problems or in redesigning products for ease of manufacture, since product life cycles have become very short because of technological breakthroughs or competitive pressures.

Another important reason for the increased attention to DFM is that global products have developed into very opposing roles: either they are commodities, with very similar features, capabilities, and specifications; or they are very focused on a market niche. In the first case, the manufacturers are competing on cost and quality, and in the second they are in race for time to market. DFM could be a very important competitive weapon in either case, for lowering cost and increasing quality; and for increasing production ramp-up to mature volumes.

Manufacturing companies have long gone past the traditional method of "throwing new products across the wall to manufacturing." There are many instances today of bringing manufacturing knowledge into the design cycle, a concept that is called *concurrent engineering.* Concurrent engineering is developed through manufacturing guidelines, manufacturability teams, and having engineers rotate through manufacturing and design areas similar to Japanese engineers.

There is a need to go beyond these first steps into a framework approach to DFM, by integrating the information generated from world-class manufacturing such as just-in-time (JIT), statistical process control (SPC), and computer integrated manufacturing (CIM) back into the design cycle.

Several tools and methods have become widely used for DFM, and will be presented and discussed with case studies and exercises. These involve quality function deployment (QFD) or customer-driven engineering for focusing the product user's input into the design specifications of new products; rating products for assembly (both manual and robotic) on the basis of part geometry or assembly motions; setting the new product specification limits to match the manufacturing process capability in terms of production average and variability; the methods of robust design for reducing product and process variability; estimating tools for low-cost fabrication of parts and assemblies using sheet metal, plastic or printed circuit boards; and finally the goals of achieving long product life through the techniques of reliability engineering.

It has been well documented that the product design cycle is the opportune time to realize the goals of low-cost, high-quality products. The design engineers should use the tools of CE and DFM to make the right decisions in terms of product, process, and material selection.

Acknowledgments

I wish to thank the people who were instrumental in encouraging and nurturing me during the long period when the book was under development. First and foremost was my family, who tolerated my mental absence during the long periods of writing and editing, especially my wife Jackie and our four children Michael, Gail, Nancy, and Jonathan. I would also like to thank the contributing authors who made the job easier by writing one-third of the chapters.

The principles of concurrent engineering discussed in this book were learned, collected, and practiced through working seventeen years at Hewlett-Packard Company and three years at the faculty of the University of Lowell, where working as a teacher, researcher, and consultant to different companies increased my personal knowledge and experience in the field of concurrent engineering.

Particular thanks goes to Dean Aldo Crugnola of the College of Engineering at the University of Lowell and Mike Critser of Cahners Exposition Group who, when asked their opinions at the start of the project, indicated their confidence and faith in my ability to finish it. In addition, my thanks to Mr. Steve Chapman of Van Nostrand Reinhold, who first approached me about the book in Anaheim, California, in February 1990, and guided me through the difficult first steps of the project; and to Dr. John McWane of Hewlett Packard, who, as a personal friend and fellow author, was instrumental in critiquing the original proposal of the project.

My special thanks go to the practitioners of concurrent engineering at major electronics companies: Mr. Happy Holden of Hewlett Packard, Mr. Anthony Martinez of IBM, and Mr. A. J. Overton of DEC, who reviewed some of the early ideas for the project and gave very useful inputs. I am also deeply grateful to two editors of VNR, Paul Sobel and Robert Marion, who were very insistent on making sure that every detail was taken care of, and to Elaine Seigal who helped edit the early rough drafts and instilled some English language discipline.

Finally, thanks to the many attenders of my seminars on the DFM and Quality Methods, including the many in-company presentations, who gave me invaluable feedback on focusing the material and clarifying the case studies and presentation of the data through the different figures and techniques.

ACKNOWLEDGMENT TO CONTRIBUTING AUTHORS

I wish to thank the co-authors who contributed five different chapters in this book. The six authors and the chapters they penned are as follows.

Mr. Masood Baig, a development engineer with Hewlett Packard Company, who wrote Chapter 9, "Geometric Dimensioning (GDM) and Tolerance Analysis," and developed the case studies for that chapter.

Dr. Dave Cinquegrana, a former student and an application consultant with ICAD, a supplier of CAD/knowledge-based design systems, who wrote Chapter 13, "Knowledge-Based Design." He developed the chapter diagrams and the software shown in the chapter.

Mr. Bill Neill, a manufacturing specialist with Hewlett Packard Company, who wrote Chapter 12, "Computer-Integrated Manufacturing (CIM); Impact on Concurrent Engineering," and coordinated the different aspects, examples, and roles of information technology in concurrent engineering.

Dr. Tim Rodgers, a development engineer with the Printed Circuit Division of Hewlett Packard Company, who wrote Chapter 10, "Design for Printed Circuit Manufacture," and collected the data and special case study for the chapter.

Dr. John Moran and Stan Marsh of GOAL/QPC, a nonprofit organization helping companies to continuously improve their quality, productivity, and competitiveness, who co-wrote Chapter 7, "Customer Driven Engineering," and developed the case study, charts, and figures from the data supplied.

Contents

Preface		v
Acknowledgments		vii
1	INTRODUCTION: DFM CONCEPTS	1
	1.1 Why concurrent engineering?	4
	1.2 Concurrent engineering as a competitive weapon	7
	1.3 Using structure charts to describe the process of concurrent engineering	8
	1.4 Concurrent engineering strategy and expected benefits to new product introduction	18
	1.5 Concurrent engineering results in the introduction of a new electronic product	19
	1.6 Conclusion	22
	Suggested reading	23
2	NEW PRODUCT DESIGN AND DEVELOPMENT PROCESS	24
	2.1 The overall product life cycle model	25
	2.2 The role of technology in product development and obsolescence	27
	2.3 The total product development process	32
	2.4 The design project phases: milestones and checkpoints	39
	2.5 Project tracking and control	42
	2.6 Conclusion	46
	Suggested reading	47
3	PRINCIPLES OF DESIGN FOR MANUFACTURING	48
	3.1 The axiomatic theory of design	49
	3.2 The design guidelines	50
	3.3 A DFM example: The IBM Proprinter	62
	3.4 Setting and measuring the design process goals	65
	3.5 Conclusion	66
	References and suggested reading	67

4	**PRODUCT SPECIFICATIONS AND MANUFACTURING PROCESS TOLERANCES**	**68**
	4.1 The definition of tolerance limits and process capability	68
	4.2 The relationship between manufacturing variability and product specifications for new products	70
	4.3 Manufacturing variability measurement and control	79
	4.4 Setting the process capability index	98
	4.5 Conclusion	100
	Suggested reading	100
5	**ORGANIZING, MANAGING, AND MEASURING CONCURRENT ENGINEERING**	**103**
	5.1 Functional roles in concurrent engineering: Design, manufacturing, marketing, quality, and sales	104
	5.2 Design guidelines	107
	5.3 Organizing for concurrent engineering	109
	5.4 Measuring concurrent engineering	115
	5.5 Conclusion	118
	Suggested reading	118
6	**ROBUST DESIGNS AND VARIABILITY REDUCTION**	**120**
	6.1 On-line and off-line quality engineering	120
	6.2 Robust design techniques	122
	6.3 Robust design tool set	128
	6.4 Use of robust methods in engineering design projects	142
	6.5 Conclusion	145
	Suggested reading	145
7	**CUSTOMER-DRIVEN ENGINEERING. QUALITY FUNCTION DEPLOYMENT**	**147**
	7.1 Introduction	147
	7.2 Quality function deployment	147
	7.3 QFD and design systems	150
	7.4 The four phases of QFD	151
	7.5 Quality function deployment case study	156
	7.6 Conclusion	184
	7.7 Glossary of QFD terms	185
	Suggested reading	186
8	**THE MANUFACTURING PROCESS AND DESIGN RATINGS**	**188**
	8.1 The manufacturing process for electronic products	189
	8.2 Design ratings for manual assembly	194

CONTENTS xiii

	8.3	Design for automation and robotics	196
	8.4	Examples of design for manufacture efficiency	204
	8.5	Conclusion	206
	Suggested reading		206
9	GEOMETRIC DIMENSIONING AND TOLERANCE ANALYSIS		208
	9.1	GDT elements and definitions	209
	9.2	Cylindrical tolerance zones	211
	9.3	Datums	213
	9.4	MMC, LMC, and RFS	216
	9.5	Controls	217
	9.6	Feature control frame	219
	9.7	Tolerance analysis	220
	9.8	Tolerance analysis case study	228
	9.9	Conclusion	233
	Suggested reading		234
10	DESIGN FOR MANUFACTURE OF PRINTED CIRCUIT BOARDS		235
	10.1	Printed circuit design	237
	10.2	DFM program requirements	242
	10.3	Performance measures	253
	10.4	Overall process	254
	10.5	Conclusion	255
	Suggested reading		257
11	RELIABILITY ENHANCEMENT MEASURES FOR DESIGN AND MANUFACTURING		258
	11.1	Product reliability systems	259
	11.2	Design tools and techniques for enhancing reliability	264
	11.3	Product testing for enhancing reliability in design and manufacturing	269
	11.4	Defect tracking in the field	273
	11.5	Summary	274
	Suggested reading		275
12	TOOLS FOR DFM: THE ROLE OF INFORMATION TECHNOLOGY IN DFM		276
	12.1	Information technology's role in DFM	277
	12.2	Information technology requirements for DFM	283
	12.3	Planning the implementation of technology to support DFM	296

12.4	Implementing DFM technology	307
12.5	Lessons learned	311
12.6	Conclusion	312
	References and suggested reading	313

13 KNOWLEDGE-BASED ENGINEERING — 315

13.1	Limitations of traditional CAD systems	315
13.2	Knowledge-based systems	319
13.3	Design example: plastic mold design	325
13.4	Summary	328
	Appendix A	330
	References and suggested reading	334

Index — 335

Concurrent Engineering and Design for Manufacture of Electronics Products

CHAPTER 1

Introduction: DFM Concepts

Concurrent engineering (CE) is defined as the earliest possible integration of the overall company's knowledge, resources, and experience in design, development, marketing, manufacturing, and sales into creating successful new products, with high quality and low cost, while meeting customer expectations. The most important result of applying CE is the shortening of the product concept, design, and development process from a serial to a parallel one.

    Products used to be handed off from one department to another, changing each time according to the needs and requirements of each. "Tossing new products across the wall to manufacturing" was a favorite expression in the early development of the electronics industry. Manufacturing would then begin the process of converting the product so as to be manufacturable, by changing part drawings and their tolerances; updating documentation such as parts lists, configurations and assembly drawings; and then reworking the tooling and renegotiating with suppliers. The next step in the process comprised the marketing and field service organizations, which report customer complaints on the product's use and performance versus the advertised specifications. Those departments would also record their own service technicians' reports about issues of warranty repairs, high defect rates of certain parts and assemblies, the difficulty in predicting failures, and performance of defective product repairs.

    In many cases companies did not learn from their product mistakes. Communication links would not be established to make the design and development departments aware of deficiencies so that they could "design them out" in the next generation of product. Worse still, many departments did not have the management skills, the resources, or the tracking systems in place to identify these deficiencies. They kept on producing new products with the same levels of customer satisfaction, quality, and cost, fixing problems again and again, while patting themselves on the back for their knowledge and experience in solving the same "déjà vu" problems for every product.

Companies attempted to improve the situation by creating post-production improvement systems. Cost-reduction programs and value engineering were introduced to formally redesign the product after it had been released to manufacturing in order to reduce cost or increase quality. Training programs were developed. Manufacturing and field service departments issued guidelines and examples of do's and don't's in the design of parts. New product development remained the domain of the research and development (R & D) department alone.

This system was shocked into action by the onset of competition from world-class companies, which moved companies towards a strategy of achieving worldwide competitive position in the manufacture of high-quality products through concurrent engineering (CE) and design for manufacture (DFM), continuous process improvements (CPI), total quality management (TQM), and "just in time" (JIT) methodology. Global competitors obviously played the new product game under different rules, and companies with traditional new product policies found themselves losing market share, profits, and even their independence, as they had to merge with or were taken over by others, or, worse still, completely disappeared. This marked the beginning of the recognition that concurrent engineering and design for manufacturing were indispensable in new product development.

Concurrent engineering is the direct involvement of all parts of an organization into the conceptualization, specification, development, manufacturing, and support of new products. The ultimate goal of concurrent engineering is to produce products that meet customer expectation, with high quality and low cost. Successful new product development is no longer the sole responsibility of the R & D department, but is the combination of efforts, teamwork, and cooperation of the total organization.

In the worldwide competitive arena, companies have to react quickly with new products in order to respond to customer trends, technological advances, and competitors' products. In addition, they need new products to continue to grow by opening new markets, creating customer demand, and increasing their market share. The key to success in these competitive strategies is the company's ability to create new products quickly by doing development, manufacturing, and delivering customer satisfaction "right the first time." It is well documented that new products are the main revenue generators for electronics companies: as their sales replace old product sales, and these new products are more profitable because they take advantage of technological changes in design and manufacturing.

Growing electronics companies are aware of the leverage of new products,

INTRODUCTION: DFM CONCEPTS

and spend a substantial part (more than 25%) of their revenues on developing new products as compared to more established companies, which spend less (5-10%). Concurrent engineering can play a significant part in the effective application of this investment: The techniques of concurrent engineering focus on the product concepts in order to meet market and customer expectations and reduce the time and iterations of new product development by producing prototypes that are made to specifications and meet the company's manufacturing requirements.

The tools and techniques of concurrent engineering were developed through the successful implementation of new ideas for engineering and manufacturing productivity in world-class companies such as Ford, AT&T, and Hewlett Packard. As these ideas have become well established, they have been codified and formalized by company trainers, academicians, and consultants into methodologies and then taught to engineers in these companies. This book focuses on collecting and explaining these concepts and ideas to engineers who are involved in new product design and development and leveraging them in developing successful new products.

When implementing concurrent engineering into an electronic company, engineers should be made aware of the values of teamwork, the sharing of ideas and goals beyond their immediate assignments and departmental loyalties. These are skills that are not taught to engineers in their formal education in U.S. technical colleges and universities, and which have to be reinforced by having these characteristics valued just as highly as the traditional engineering attributes of technical competence and creativity. The successful new product interdisciplinary teams are the ones that are focused on aggressive but achievable goals for concurrent engineering and design for manufacture. The characteristics of teamwork and cooperation can be rewarded by making them an integral part of the performance evaluation process for engineers.

One of the most important elements in understanding the complexities of introducing new products is the use of the tools of structured analysis to describe the different processes and information flow inherent in a complex electronics factory. The structure charts methodology, which was developed for the software industry, has proven to be very effective in describing and clarifying these processes. Structure charts will be presented in detail with examples in this chapter.

In this chapter, the methods and techniques of concurrent engineering will be presented as a prelude to more in-depth discussions in the following chapters. A case study, from the author's personal experience in the mid-1980s, will be used as illustration of a typical concurrent engineering effort.

1.1 WHY CONCURRENT ENGINEERING?

There have been many discussions about the competitive posture of the U.S. economy, and in particular, certain segments that faced competition from abroad. Books, articles, and academic research have focused on the disparity of U.S. and foreign products in product development costs and cycle times, manufacturing cost and quality, reliability and customer satisfaction. Many U.S. companies have sent engineers and managers to study these foreign successes in order to try to unlock this mystery. What they found was 99 per cent common sense, and the old adage was rediscovered: "Work smarter, not harder."

A look at the auto industry (see Table 1.1) illustrates the main advantages of concurrent engineering and design for manufacture. As can be seen in Table 1.1, the ratio of the two measures is about half for the Japanese as compared to the U.S. auto industry. Given that the engineering universities and institutions in Japan are not much different from those of the United States, one has to look elsewhere for reasons for these differences.

The results of concurrent engineering are shown in Table 1.1. Since the figures for the design time and effort of Japanese cars are roughly half of those for the United States, the Japanese can support a smaller volume and lifetime of each product and a larger number of different products in the marketplace. This competitive advantage allows for a successful sales strategy and therefore larger overall profit.

TABLE 1.1. Japanese/U.S. Auto Design and Product Cycles

	Japan	U.S.
Design time per model (months)	47	60
Design effort per model (man-hours)	1.7	3.0
Average replacement period per model (years)	4.2	9.2
Average annual production per model	120,000	230,000
Models in production	72	36

© *The Economist* Newspaper Limited, April 14, 1990. Reprinted with permission.

WHY CONCURRENT ENGINEERING

With so many products, the market can be partitioned appropriately, with different products focused on different segments of the market. In addition, since the majority of the profits occur in the early part of the cycle for successful products, the products are turned over much faster, allowing them to be retired at close to the optimum profitability. Before these products mature and their sales and profits decrease, they are replaced by newer products, which are even more responsive to customer demands, with the advantages of newer and lower-cost technology.

It is apparent in this environment of shorter overall life cycles, both in product development and in the market place, that the new products have to hit the market with customer expectations correctly set and with a high level of quality, and be priced competitively. There is clearly no time to correct design mistakes and errors and to reengineer the product for lower cost or higher quality. The perceived customer loyalty will also suffer if the product is released prematurely, resulting in excessive changes and recalls. With the help of concurrent engineering, new products have to meet customers' needs in a timely fashion and with low cost and high quality.

Concurrent engineering is the answer to the need for shorter development cycles: New products are designed with inputs from all concerned. First and foremost is the customer, who will determine the ultimate success of the product. If the target customer requirements are well defined and documented, then the product specifications can be focused on customer needs. The methods of quality function deployment (QFD) are designed to listen to the voice of the customer, especially for evolutionary products, where the customer is well aware of the current choices and capabilities of available products.

For the manufacturing department, concurrent engineering means the development of new product specifications and manufacturing tolerances for the lowest production costs and highest-quality products. The production rate can be ramped up to full mature volumes very quickly after release to production, since the manufacturing process is well documented and controlled using statistical process control and other total quality management (TQM) tools.

Concurrent engineering means that other interested departments such as field service, quality, marketing, and manufacturing contribute strongly to the new product design and specifications, to assure good warranty levels, easy serviceability, and testability of the product. These departments, which historically were not considered or consulted in the design process, become an important part of the product goal-setting process, with measurable and quantifiable targets.

1.1.1 The Leveraged Effect of the Design Cycle on Life Cycle Costs

In many industries, there is increased understanding of the importance of the correct decisions being made at the earliest possible time in the product life cycle, specifically at the concept and development phases. These phases represent the smallest amount of overall product cost expenditures, but the effect of the right decisions made at these stages is highly leveraged, because they have the greatest impact on the overall life cycle costs. These leveraged effects are illustrated in Figure 1.1. More recent data verifies this concept, as shown in Table 1.2.

The important lesson to be learned in new product development is to use concurrent engineering methods to design the products *right the first time*, using inputs from marketing, manufacturing, quality, and field service. This book will illustrate some of the tools and techniques available to quantify, implement, and measure the progress of the new product development teams towards achieving these goals.

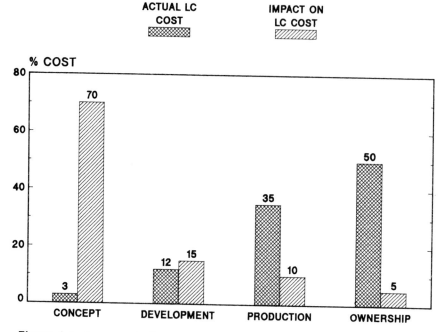

Figure 1.1. Leveraged effect of the design phase on life cycle costs. (*Source: Military Electronics*, August 1980)

TABLE 1.2. Leveraged Effect of Design Phase in the Electronics Industry

(a) With design decisions so critical, design is a tiny piece of the development pie, but it locks in the bulk of later spending

Life cycle phase	Percentage of total cost (Cumulative)	
	Incurred	Committed
Conception	3–5	40–60
Design	5–8	60–80
Testing	8–10	80–90
Process planning	10–15	90–95
Production	15–100	95–100

(b) With engineering so expensive, the cost of an engineering change increases logarithmically when done later in the product cycle

Time design changes made	Cost
During design	$1,000
During design testing	$10,000
During process planning	$100,000
During pilot production	$1,000,000
During final production	$10,000,000

Source: Business Week, April 30, 1990.

1.2 CONCURRENT ENGINEERING AS A COMPETITIVE WEAPON

One of the results of concurrent engineering is that there is no plateau a company can reach in the rate of decreasing development time or in turning production around. Table 1.3 illustrates the point.

As soon as a company accomplishes a significant increase in product development or in manufacturing turnaround of orders, the competition is probably gearing to go further down the curve in terms of faster development and manufacturing response time. It is not sufficient to set goals in today's competitive world that are evolutionary. A 10% improvement is not adequate. Companies today, using the tools of concurrent engineering, are able to implement improvements of significant orders of magnitude in quality, timeliness, and costs of new products. For example, in the 1980s, Hewlett

TABLE 1.3. Time-Based Capabilities are Becoming More Important

	Fast Innovators		
		Development time[a]	
Company	Product	Old	New
Honda	Cars	5 y	3 y
AT&T	Phones	2 y	1 y
Navistar	Trucks	5 y	2.5 y
HP	Printers	4.5 y	2 y

	Fast Producers		
		Order to ship[a]	
Company	Product	Old	New
GE	Circuit breakers	3 wk	3 d
Motorola	Pagers	3 wk	2 h
Brunswick	Fishing reels	3 wk	1 w
HP	Test equipment	4 wk	5 d

© *Fortune* magazine, February 13, 1989. Reprinted with permission.
[a] y = year; wk = week; d = day.

Packard instituted a program to improve the quality of its products by a 100% in five years.

Mercury Computers is a small company based in Lowell, Massachusetts, that produces speciality personal computer-based computational products. It has been able to shorten the cycle of shipping printed circuit boards (PCBs), from design approval to customers' hands, to less than 90 days. The significance of this is greater because a mix of in-house and outside suppliers were used to lay out, fabricate, assemble, and test the PCBs. This effort included a wide mix of engineers, equipment, and technologies at the suppliers' and the company's facilities. The company bases its success on the understanding of its customers' needs as well as the manufacturing and field processes.

1.3 USING STRUCTURE CHARTS TO DESCRIBE THE PROCESS OF CONCURRENT ENGINEERING

Structured analysis (SA) and structured designs are tools that were developed for software development as means of a hierarchical decomposition and description of software modules. Structured analysis and design were developed to replace the traditional tools of flowcharting as the magnitude of software projects and programming complexities increased.

Another advantage of structured analysis and design is the presentation of information flows between different systems and departments in a graphical manner, which together with the hierarchical approach, allows for easy understanding of a complex system or process. Structured analysis and design have been used successfully in management information systems to design the information and data for a detailed manufacturing operation. In this manner, it can be used to describe the complex marketing, sales, manufacturing, and quality systems that are used to develop and introduce new products to manufacturing and the marketplace.

Today, structured analysis has been more formalized for specific applications, and there are many software systems available in the marketplace to assist in the development and the drawing of structure charts. The basic elegant simplicity of structured analysis is that it uses only few symbols and techniques to present a complex system or operation. The top-level boundary of the system being described is called the *context diagram*, and the decomposition of the system into smaller more detailed units involves *data flow diagrams*. This process, known as "top down partitioning" or "leveling," occurs where data flow diagrams are decomposed from a very high-level and general view of the system to a very low-level and detailed view of specific operations.

A data flow diagram may be defined as a network of related functions showing all data interfaces between its elements. These elements are as follows.

- The data *source* or *destination*, represented by a rectangular box. A source is defined as an originator of data and a destination is defined as a terminal target for data. Sources and destinations are used to determine the domain of study of the system. In concurrent engineering, they refer to departments, suppliers, and customers.
- The *data store* is represented by two parallel lines or an open box. It represents a repository of information. Data can be stored in electronic files or in physical locations such as file drawers. The name of the file or the storage system should be written alongside the symbols. In complex diagrams, the same data stores might be drawn at different locations to minimize nesting of the lines. In these cases, another vertical line is added to the symbol to indicate that it is repeated in the diagram.
- The *data flow*, represented by an arrow, symbolizes the information being carried from different parts of the systems in order to be transformed or recorded. The name of the piece of data should be written alongside the arrow.

Every data flow and data store should have an associated *data dictionary*, which provides a single document to record information necessary to the understanding of the data flow diagrams. The information can take the form of what records are kept for each data item and the associated information for each record.

An example of a data dictionary is shown below

Data dictionary example

Part master file	= Part number
	+ Description
	+ Supplier
	+ Supplier part number
	+ Cost
	+ Lead time
	+ Order quantity
	+ Balance on hand
	+ Incoming inspection code
	+ Shelf-life
	+ Storage location
	+ Where used + ... +
Part structure file	= Part/assembly/fabrication #
	+ Description
	+ Quantity used
Where used file	= Part/assembly/fabrication #
	+ Description
	+ Parent number

In this example three files that are commonly used to document all parts, fabrication items, and subassemblies in a manufacturing enterprise are outlined:

Part master file, where the attributes of all part numbers in a new product, such as part number, description, suppliers, cost, lead times, inventory levels, and other manufacturing information, are kept.

Part structure file, where all parts, fabrication items, and subassemblies necessary to make the total product are recorded by their relationship to each other, in a top-down approach. Each one is recorded by the list of items required to make it.

Where used file, which is an inverted part file to identify the parent part number or assembly for each part.

The process, is represented by a circle. A process transforms the data flowing in the system either by reformatting the data, or by appending or deleting parts of the information; or the process can use different sources of data to generate new information. Partitioning of the data flow diagrams into more detailed lower levels should be terminated at a point where the tasks of each process are easily understood in their entirety. A process that can no longer be decomposed without the resulting processes appearing trivial is called a *primitive process*.

At this level, where a process cannot be decomposed further in functionality, it should have a written *process specification* (also known as a mini-spec or P-spec) outlining the name of the process, its inputs and outputs, and the process specification on data transformation or generation. Commonly used tools for describing the process specifications involve symbolic methods such as decision trees and decision tables. These tools record process decision making based on input flows in a shorthand manner. For more detailed information on these tools, see some of the reference materials.

An example showing the process specification for handling a part shortage in a stock room could be as follows:

When balance on hand is smaller than requisition, then

    Issue remaining parts against requisition
    Alert purchasing agent
    Fill out a shortage alert form
    Tag the stock room bin with a red shortage tag

These elements of structured analysis, shown in Figure 1.2, are very useful in documenting and explaining to the R&D department the methods, techniques, responsibilities, and operations of the different parts of the organization. It serves as a powerful documenting tool for the current process. In addition, the inherent inefficiencies of the process will be visualized easily so that it can be optimized by eliminating excess loops and transcription of data.

Each department should record its procedures, responsibilities, and functions in its own data flow diagram. This serves as an excellent documentation tool for the total process and its interactions. The visual presentation of the diagrams is much easier to comprehend than written procedures and documentations. New product design engineers can quickly grasp the interconnection of the different parts of the organization in producing prototypes and presenting data.

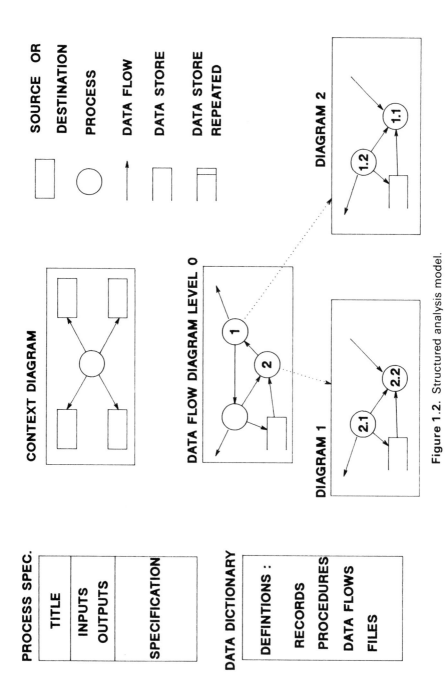

Figure 1.2. Structured analysis model.

USING STRUCTURE CHARTS 13

1.3.1 Case Study: Structured Analysis to Quantify a Production System

In this example, illustrated in Figures 1.3 through 1.6, structured analysis was performed to describe the production system in an electronics factory. An interdisciplinary team was formed to recommend a new production operational strategy, and team members quickly found themselves unaware of how the different departments in their plant carried on their tasks and interacted with other departments.

Using the tools of structured analysis and data flow diagrams, the team members collectively produced the context diagram and the top-level (called 0 level) data flow diagram. Each team member then graphically presented their individual department's processes and interfaces by developing their own department's data flow diagram, which was in synchronization with the top-level diagram. The charts were helpful for team members in understanding the overall manufacturing processes and their interactions, and were used as the basis for formulating a new strategy for the production function of the company.

The diagrams contain data stores that are named by acronyms particular to the manufacturing operation. Their intent was to generally qualify the manufacturing process flows and documentation, and not to specifically detail every existing operation and process. Although no data dictionaries or process specifications were provided for this current process, the reader can follow the information and data flows through the different departments and understand the complexity and interconnection of the different systems involved in procuring, inventorying, producing, and testing the product. When designing new manufacturing processes, it is advisable to create the data dictionaries and process specifications to identify each procedure exactly.

The data flow diagrams can be used as a quick reference to engineers and managers in understanding and following the manufacturing system procedures and requirements. They can lead to better management of the manufacturing function and the data structure need to support it by providing a visual representation of the connectivity of the different departments, databases, and functions to be performed. The results of charting manufacturing into data flow diagrams will result in a well-managed and efficient operation by:

- Eliminating redundant operations, which will become apparent once the total process is visualized.
- Improving the efficiency of existing operations, by clearly identifying the responsibilities of each one and its relationship to other operations, as well as the information necessary for correctly performing its functions.
- Better integration with outside activities and sharing of existing resources rather than developing new ones, as the diagrams provide a

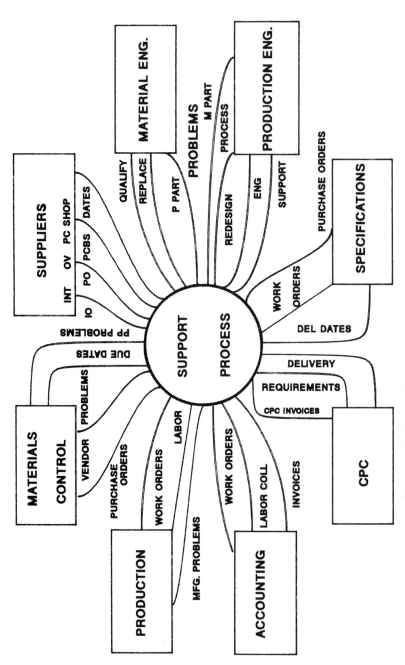

Figure 1.3. Production support process context diagram.

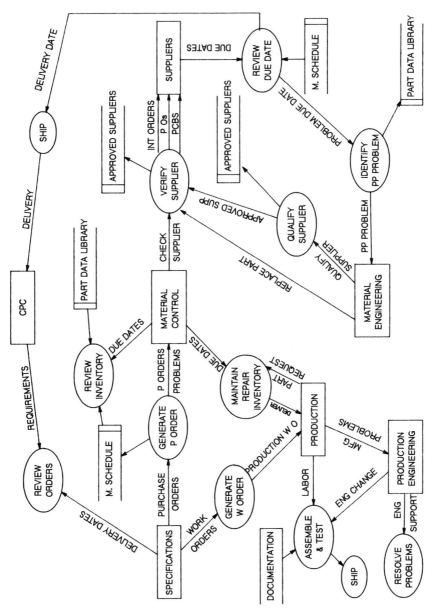

Figure 1.4. Production support process data flow diagram (0).

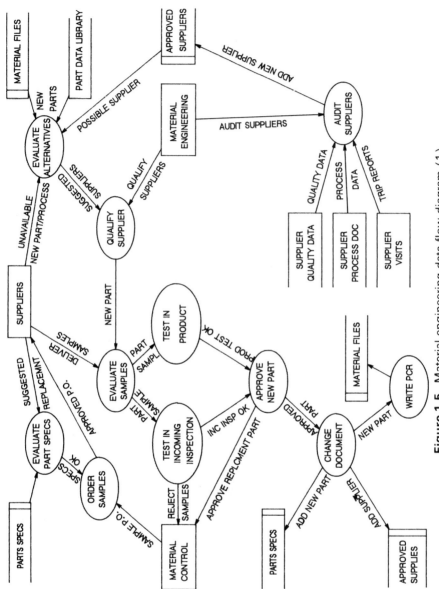

Figure 1.5. Material engineering data flow diagram (1).

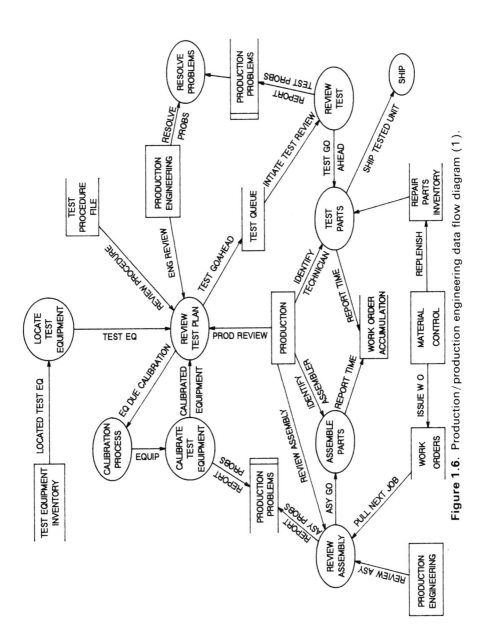

Figure 1.6. Production/production engineering data flow diagram (1).

ready description of the procedures and documentations of the current process.
- Increasing data integrity by eliminating excess and redundant operations. More accuracy will result when databases are well connected, consulted more frequently, and used in more applications. With more focused attention, data has a greater chance of being maintained correctly.

1.4 CONCURRENT ENGINEERING STRATEGY AND EXPECTED BENEFITS TO NEW PRODUCT INTRODUCTION

The first concurrent engineering effort in many companies is the formation of an interfunctional or interdisciplinary team to facilitate new product development. The purpose of the team is to shorten the product development cycle, improve product reliability, and reduce cost through concurrent engineering and design for manufacture. At this stage, it is desirable to set specific goals for the effort. These goals should be aggressive but realizable for the company and should be based on the performance of current products.

The strategy of the concurrent engineering effort is to combine the connectivity of CAE/CAD/CIM with the design for manufacturing and to get manufacturing involved early in the design of the product. Elements of the strategy could be: (1) Document the current manufacturing process capabilities and constraints. Structured Analysis and data flow diagrams should be generated for the production processes and reviewed by the responsible departments for elimination of redundant tasks and verification of the current processes. (2) Review all new parts and assemblies for manufacturability, serviceability, testability, and repairability as well as the traditional focus of design reviews such as technical correctness and overall design quality. (3) Develop the computer integrated network connectivity for downloading CAE/CAD designs into the manufacturing equipment and processes. This would eliminate the transcription errors in tooling-up the fabrication and assembly processes such as printed circuit board fabrication, assembly, test, and sheet metal enclosures. (4) Use and develop software tooling whenever possible, to allow manufacturing to build all prototypes. Manufacturing should treat prototypes as the highest priority of production. When making prototypes, manufacturing should record all discrepancies in assembly and fabrication instructions, violations of manufacturing equipment or process constraints, documentation errors, parts misalignment, and all parts failures. This manufacturability feedback should be supplied promptly to the design department, together with the prototypes, with suggested design changes. Further resolution of these issues should be done formally in manufacturability design reviews.

Concerns that the prototype priority scheduling could disturb the shop expediting process should be addressed. Usually the prototype activity should be a small percentage (less than 5%) of the overall production schedule in order to minimize the expediting in the production shop scheduling. Manufacturing should be aware that prototypes are not as well documented as current production runs, because the design engineers are reluctant to spend too much time further documenting prototypes that will undergo further changes. These concerns should not detract from the overall benefit of the design for manufacturing.

When production will be performed by an outside supplier, or the manufacturing operation is located away from the design and development department, the focus on process documentation will be much higher for all prototypes. In any case, the use of specialized departments for producing prototypes, such as the use of model shops, should be discouraged. These departments are specially staffed with the highest-skilled production personnel that can fabricate or assemble parts with incomplete information, or are capable of making prototypes to specification from marginal manufacturing processes. In these conditions, designers assume that parts and assemblies are designed to be easily manufacturable, yet when production quantities are requested from the regular production sources, parts cannot be made to current documentation or specifications.

1.5 CONCURRENT ENGINEERING RESULTS IN THE INTRODUCTION OF A NEW ELECTRONIC PRODUCT

Typical results from this initial effort of concurrent engineering can be shown in improvements to the development process metrics. These improvements are based on the performance of older products in the development cycles, and therefore have to be normalized by component counts or percentage of first-pass yields. Typical charts are shown in Figures 1.7 through 1.10. These charts show the increased efficiency of new product designs, in terms of the following:

1. Shorter time of the product introduction cycle, as indicated in production ramp-up to mature volumes (Figure 1.7). This is very important in order to gain maximum revenue from the new product, since the date of obsolescence of the product is fixed in time due to competitive and technology enhancement (see Chapter 2).

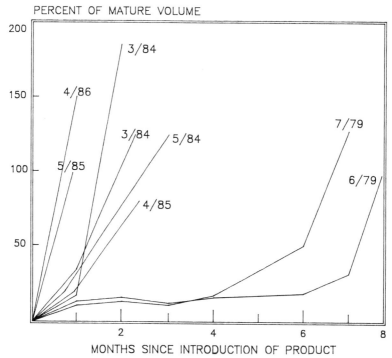

Figure 1.7. Shortened new product introduction cycle.

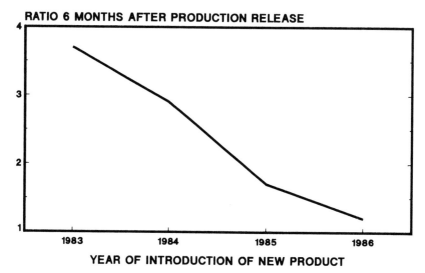

Figure 1.8. Improved design quality for new products. Ratio = (total engineering changes on released assemblies) / (total number of released assemblies).

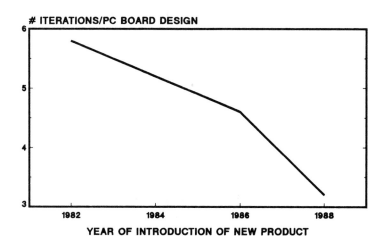

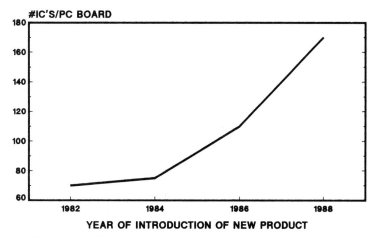

Figure 1.9. Reduction of design iterations for new products.

2. Improved design quality, as measured by the amount of engineering changes made in the early part of the product life, 6 months after release (Figure 1.8). These engineering changes measure the design robustness of the product and the completeness of documentation. The data is normalized to the total number of released assemblies.
3. Reduction in design iterations, taking into account the amount of design complexity (Figure 1.9). In electrical designs that are implemented in printed circuit boards, the number of components is a good indicator of the design complexity. Design iterations should continue to decline even as the design complexity increases. The acquisition of new

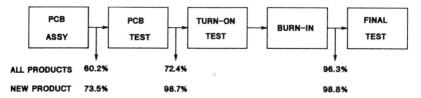

Figure 1.10. Increase in manufacturing process quality for new products.

technology in terms of computer aided (CAE) can also benefit this trend by allowing for analysis and verification of the design, as well as automating parts of the drafting, such as automatic folding for sheet metal and automatic routing for printed circuit boards.

4. Increased process quality, as measured by the early production process yields for the new product parts and assemblies. For printed circuit boards (PCBs), these yields are at the test point at PCB assembly and final product assembly and test (Figure 1.10). Typically, the source of defects for PCB assembly are missing, misloaded and reversed components; soldering defects; and defective incoming components. The operational strategy for the management of these production processes has to be closely connected to the development goals of the new products. The quality and productivity levels of these operations in terms of equipment plans and acquisitions, operator motivation and training, and long-term technology strategy have to be closely coupled to the new product introduction schedules. Good process capability documentation and feedback of early prototypes to the designers can greatly influence the level of assembly quality.

5. Reduced production times, where the new product assembly and fabrication operations should achieve mature production times at the earliest possible production runs. Typically, there is a learning cycle, where operators start with high production times per unit, then reduce the time as they learn how to make the product once the documentation and manufacturing processes are established. This phenomenon is called the *learning curve*, and it should not exist for new products developed using concurrent engineering.

1.6 CONCLUSION

Concurrent engineering encompasses all the elements of speeding up the conceptualization, development and manufacturing of new products. The tools and techniques to be discussed in this book have been used by world-class manufacturing companies to produce new products for customer needs with high quality and low cost. It should be apparent to all that concurrent engineering is a requisite for companies developing new products, and must be used to develop aggressive but achievable goals of improving new product quality at lower costs, and with high serviceability and customer satisfaction.

One of the techniques presented in this chapter was structured analysis and data flow diagrams for documenting the overall product development process.

Typically, concurrent engineering is introduced to an electronics company in terms of assembling an interdisciplinary team to pool their knowledge in designing new products. The techniques, tools, and methodologies of concurrent engineering are meant to augment the traditional R&D development process in terms of making it more responsive to customer needs.

SUGGESTED READING

Cutts, L. *Structured Analysis and Design Methods*. New York: Van Nostrand Reinhold, 1990.

Demarco, Tom. *Structured Analysis and Systems Specification*. New York: Yourdon Press, 1978.

Fitzgerald, J. *Fundamentals of Systems Analysis*. New York: Wiley, 1986.

Gane, C. and T. Sarson. *Structured Systems Analysis: Tools and Techniques*. Engelwood Cliffs, N.J.: Prentice-Hall, 1979.

Myers, Glenford, J. *Composite/Structured Design*. New York: Van Nostrand Reinhold, 1978.

Page-Jones, Meilir. *The Practical Guide to Structured Systems Design*. New York: Yourdon Press, 1980.

Pugh, Stuart. "Engineering design: Unscrambling the research issues." *Journal of Engineering Design*, Vol. 1, No. 1, 1990, pp. 65–72.

Shina, S. "Benefits of concurrent product process development." *HP Corporation Executive Seminars*. September, December 1986.

Stevens, Wayne P. *Using Structured Design*. New York: Wiley, 1981.

Weinburg, Victor. *Structured Analysis*. New York: Yourdon Press, 1980.

Woodruff, D. and Phillips, S. "A smarter way to manufacture: How concurrent engineering can invigorate American industry." *Business Week*, April 30, 1990.

Yourdon, Edward and Constantine, Larry. *Structured Design*, 2d edn. New York: Yourdon Press, 1978.

CHAPTER 2

New Product Design and Development Process

The new product development cycle has been undergoing changes that are accelerating rapidly because of the increases in technology and competitive pressures that are rendering new products very critical to the financial future of companies and making older products obsolete much faster today. In this chapter, we will discuss the evolution of the new product development process and its implementation within the overall product life cycle.

In the electronics industry, the life cycle of the typical product is different from the historical models because of the impact of technological change. The investment in new product development is much higher than in other industries, because the rate of new product development and technological change is much higher. These development costs tend to be highest during the startup phase and lowest as the products mature. Costs vary from about 5% of revenues in established products, to more than 25% of revenue in new market segments.

Companies are looking for ways to optimize and preserve their investments. They have targeted product development as the most cost-effective stage in lowering their overall new product costs. This chapter will discuss methods and techniques for assuring success in new electronic product strategies, concepts, and development cycles.

The new product development cycle will be presented with a simplified model that includes the concept and development stages. The control and measures of the development stage will be outlined, using project phases and checkpoints. In addition, project management techniques, whether time- or event-based, as well as tools for keeping track of project status and return on investments, will be shown.

2.1 THE OVERALL PRODUCT LIFE CYCLE MODEL

Products go through many stages in their lifecycle (see Figure 2.1). The first stage is called *startup* or *market development*. During this stage, emphasis is on the performance of the product. Features such as speed, capacity, response time, and other "bells and whistles" dominate the product cycles. At this stage, the benefit of the product to the customer is perceived to be very high in increased productivity or personal comfort and satisfaction. The number of competitors is large, since entry into the market is wide open, and a new company can establish a "niche" in the market place for a relatively low investment. Product development during the startup stage is marked by the intense drive to arrive at the market as early as possible, with minimum concern over manufacturing cost. A good indicator of this stage is the number of wire cuts and changes to printed circuit boards in new products. The quality and reliability of the new product in manufacturing is achieved by extensive inspection and test.

The second stage is the *growth* stage. As the market place is expanded and general acceptance of the product is assured, the number of competitors drops and the rate of market development begins to slow. The issue is not the acceptance of the technology or the particular use of the product, but the differentiating aspects of the manufacturers. Elements of the long-term cost of ownership of the product such as the quality and field support of the product, the commitment of the manufacturers to the particular business segment, the growth of ancillary products and services supporting the product and its technology, and the increasing confidence of the customer in the evolution of the product are emphasized.

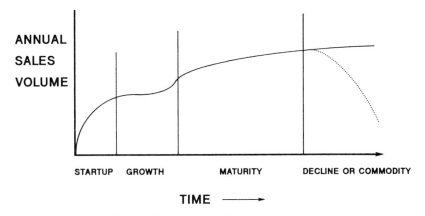

Figure 2.1. Product life cycle stages.

Product development during the growth stage is characterized by the focus on introducing the manufacturing guidelines of capabilities and constraints to the new product, and beginning to concentrate on manufacturing as a strategic weapon to emphasize low cost and high quality. Working with suppliers is increased, with the introduction of just-in-time (JIT) schedules into the manufacturing process.

The third stage is the *maturity* period. This phase is characterized by the emergence of a dominant technology or technique for the product design. At the same time, the relative growth of the market is slowed, being only proportional to the growth of the population or the customer base, as the product saturates the market. The number of manufacturers continues to decrease, as they either go out of business or are bought out by the larger companies. The competitive emphasis in this stage is on price and quality, as the dominant technology does not allow too much variation on the basic design of the product.

Product development in the maturity phase is focused on continued improvement in manufacturing process such as a stronger emphasis on quality through the tools of control charts, continuous quality improvement, and robust processes. Variability reduction through implementing the techniques of design of experiments (DOE) and heavy emphasis on automating part or all of the manufacturing processes is increased. The suppliers of parts/raw materials for the product are involved early and often, and the design process is more robust through the use of analysis and simulation tools.

The last stage can take one of two forms: Either the product will *decline* as the need for the product is overwhelmed by new technology (as were eight-track cassette players and electric typewriters), or the product will develop into a *commodity*. In either case, the number of product manufacturers will decline to a select few big companies, and entry into this market will become very expensive and risky. The emergence of standards of use, manufacture, interconnection, and quality will make price the only competitive leverage. The products will essentially be interchangeable from one manufacturer to another, with high customer expectations of quality and reliability. The revenue per unit decreases rapidly, as manufacturing techniques become the major factor in ensuring the long-term survival of the product's manufacturing company. Follow-on products will be evolutionary, with a market leader establishing a very careful trend that locks on his customer base and provides a definite upgrade path for the new generation of products.

The product development emphasis in the commodity stage is on reducing manufacturing cost while maintaining the high quality expected by the customer. There is a much higher level of automation as manufacturing knowledge and the stability of the design are increased. Few companies can enter into a market at the commodity stage as the costs are prohibitively high for recruiting the expertise required to develop the poduct at a high level of production, automation, and quality.

The electronics industry has followed many other industries into this pattern. The automobile industry is a prime example. In the early part of this century, there were hundreds of auto manufacturers, and any of the competing technologies could have become dominant: electric, steam, or internal combustion. The computer industry has gone through these stages for various products. While mainframes are in decline, personal computers have become a commodity industry, with exchangeable software programs and plug-in cards.

2.2 THE ROLE OF TECHNOLOGY IN PRODUCT DEVELOPMENT AND OBSOLESCENCE

The technology changes in the electronics industry stem from many different sources: faster and larger capacity of electronic components, such as memories and microprocessors, new techniques in information storage, retrieval and display, and new materials and processes in fabrication such as printed circuit boards and plastics. In addition, global competition is forcing the turnover of products at a much higher rate. New products quickly become old products as manufacturers throughout the world rush to produce very similar products through reverse-engineering techniques.

Traditionally, the most revenue from new products occurs while they are early in the product life cycle. If the product offers unique features that are not duplicated by the competition, then the revenue from this product will be even greater. As the product matures, and there are many competitors in the market offering similar capability, price erosion will ensue as more companies compete for scarce customers.

Typical price performance curves have been developed for electronic products (Figure 2.2). They show the price coming down for the same performance/capability over time, as competitive pressures and technological advantages take hold. At the same time, new performance curves tend to form at the top price, and then continue to come down to the lower price over time. The performance improvement for the creation of a new performance curve has to be significant: More than 30–50% improvement in speed, measurement capability, or response time is needed to form the new curve. Customers will not support a new pricing structure for the higher-performance product unless there is a measured improvement to their productivity through the use of the product.

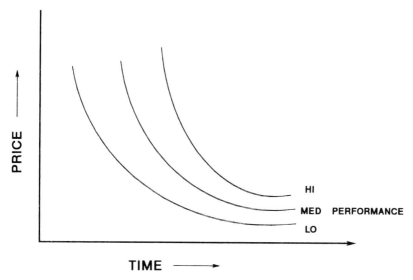

Figure 2.2. Typical price performance curves.

The application of technology trends to product development can take two different paths, depending on the new product strategy for the company. Some have opted for both paths.

- *New capability or more performance for the same price (Figure 2.3).* The advantages of this scenario are the creation of new market demand for the new technology, and providing a customer upgrade and growth path into the product as customers become more sophisticated in their needs. The pricing mechanism is easier to set for maintaining the current level of profitability, since the new products are priced similarly to older ones.

The risk of this technology strategy is that competitors can easily upset it by offering the same or increased performance at a lower price. In addition, this strategy requires a well-developed market and sophisticated customers who can understand the nuances of the price performance curves.

- *Provision of existing capability or performance for lower price (Figure 2.4).* This strategy can be either a defensive or an offensive move: It has the advantage of protecting the low-end customer base and could broaden the appeal of the product to a larger constituency. In addition, it leverages off the market development cost associated with new products.

One of the consequences of this strategy is the increase of product sales volume because of lower product selling price. However, if the sales volume

THE ROLE OF TECHNOLOGY

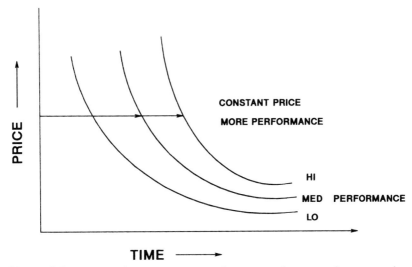

Figure 2.3. Price performance strategy 1: more performance for same price.

does not increase substantially to offset the lower sales price, profitability will be decreased. This could, in the long term, erode the financial base of the company.

Another risk to this strategy is the possible existence of hidden costs that can crop up when volume increases. The manufacturing process has to be

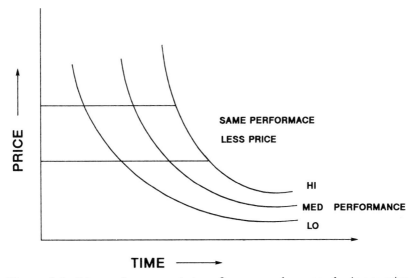

Figure 2.4. Price performance strategy 2: same performance for lower price.

well developed in documentation and employee knowhow and the product material and labor costs should be well defined.

A further concern when developing a lower-cost strategy is that lowering the price may not generate more customer sales at the expense of the competition. Surveys have shown that the effects of flexibility, perceived quality, and quick delivery are becoming more important than price in the customers' purchasing decision (Boston University Manufacturing Futures Survey, 1988).

An important role of technology on product development and strategy is the effect of technological change in obsoleting existing electronic components and processes, thereby shortening the life span of electronic products. A case study in point is the obsolescence of electronic components by a major company. In 1987, the rate of obsolescence of several electronic components was calculated on the basis of the number of obsoleted parts in the corporate database. The results are plotted on Figure 2.5. The average obsolescence rate of all parts was 7% per year. Semiconductor memories had the highest rate at 15% per year. Figures 2.6 and 2.7 illustrate this point by looking at memory lifetime trends. This obsolescence can be negated by making lifetime buys, a dangerous proposition, considering there might be shelf-time issues in the storage of these electronic components, in terms of soldering and assembly.

These technological factors and competitive pressures determine the time of obsolescence of the product. Many manufacturers hold on to products

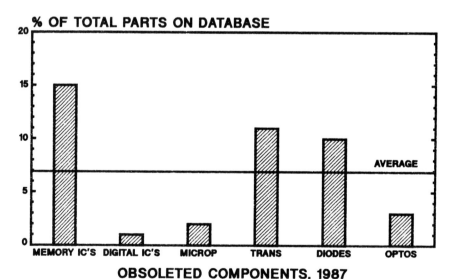

Figure 2.5. Obsoleted electronic components, 1987.

THE ROLE OF TECHNOLOGY 31

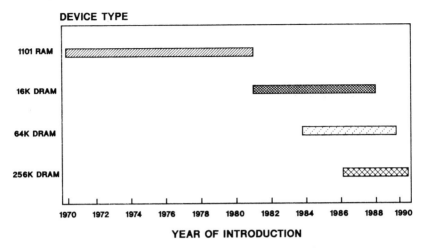

Figure 2.6. Memory devices lifetime by device type.

too long, only to discover that the negative effects far outweigh the easy profit they think they can generate: Customers may feel that these companies are lagging in the technological field, and will not purchase these products; the manufacturing department will have to incur the extra expense of procuring and maintaining outdated parts and processes; there is a loss of the professional staff morale, as engineers struggle with maintaining old technologies.

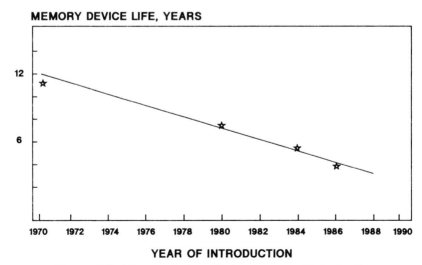

Figure 2.7. Memory devices lifetime by year of introduction.

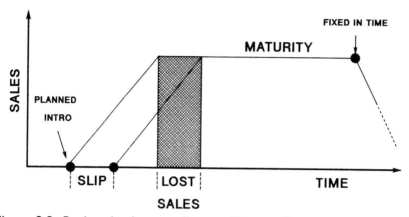

Figure 2.8. Product development slippage effects on life cycle sales. Note that a 1-month slip in product development means one less month of mature sales.

These factors emphasize the importance of designing products on time and quickly ramping up production to mature levels. This is especially true since the end-point of the product life cycle is fixed in time because of technological issues beyond the control of the company. Figure 2.8 illustrates this point. Any slip in release to production, or an early production problem, translates to lost sales that are not recovered at the end of the product life cycle.

Another issue related to being late in the design phase is the effect of releasing a new product on the current sales of existing products. If the new release plan does not dovetail perfectly with the overall production schedule, the company could be vulnerable to a dry period, where customers convert their orders from old products to new ones as soon as the new products are announced. This could create a serious gap in the revenue stream and production schedules if the new products have late shipment because of production problems. The production turnover plan has to be very smooth to carry the company financially during the product changeover.

The classic case of this type of dilemma concerns the Osborne computer in the early 1970s. The company was a small startup, with a unique idea (at that time) of having a personal computer in a briefcase. The Osborne I was very successful, but when plans were announced for the Osborne II, customer orders shifted to the newer model. The company was unable to ship the Osborne II on time, and eventually went out of business.

2.3 THE TOTAL PRODUCT DEVELOPMENT PROCESS

New products are developed in companies based on the long-range strategic and business plans for the market segment in which the company wants to operate. The new product strategy plan is very dependent on the product/market life cycle phase: Whether the product is in the startup, growth, mature, or commodity phase will greatly influence the capability and performance versus price range versus timing of the product introduction.

To formulate a new product strategy that is cohesive with the rest of the entity, it is important to begin with the company goals. They should outline which business segment is targeted and, hopefully, the boundaries of that segment. In addition, they should define the competitive advantage of the company; for example, innovation and technology, cost and manufacturing technology, quality effort, customer satisfaction, and organizational flexibility.

The component of the product strategy should include a hierarchy of elements that define the strategy over time:

Mission statement—A broad statement for the next 10–20 years.

Intermediate range plan—A more detailed plan outlining goals and actions to be taken for the next 3–10 years.

Product plan—Products (performance and price) to be introduced for the next 1–5 years

Tactical plan—Action plan for the short range of 12–18 months to accomplish objectives

The new product development process varies with different companies, depending on market requirements, competitive pressure, and the internal company strategy and methodology. Most companies have "go/no-go" decision points at various stages of the concepts through the development stages, with checkpoint meetings and targets to be met and revisited.

Figure 2.9 is a simplified flow diagram of the new product development process. It divides the process into two major phases: concept and development. There is only one "go/no-go" decision point, which occurs when deciding to go from the concept to development phase.

The concept stage begins with the identification or the creation of a specific project or product team. The product idea can develop from several sources: competitive evaluations, marketing and customer surveys, and most importantly, from the *product champion*. The product champion, whether he or she is a high-ranking executive or a staff engineer on the bench, plays a very important part in the introduction of new products. He or she has the vision, the focus and the determination to carry through his or her new product ideas into fruition. His or her risks are high but the rewards can be great. If the product champion's ideas are not implemented, he or she might leave the company to start up new ventures. Apple Computer was developed partly by an engineer who could not convince the management of Hewlett Packard Corporation to act on his ideas for personal computers.

Once the product is identified, an iterative process begins to take shape in order to further refine, clarify, and specify the product definition. This

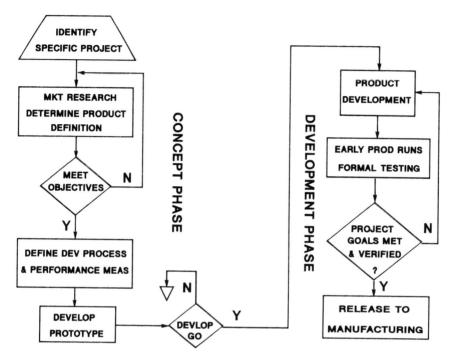

Figure 2.9. New product development process.

phase is enhanced by market research, surveys, and economic justifications. Once the process is completed, the development and the performance measures of the product are identified in terms of product specifications, potential revenues, costs, product life cycle and impact on existing company products.

The results of this iterative process is sometimes built into a *business plan*. The plan is the blueprint for developing, manufacturing, and selling a new product to the marketplace. It is imperative for all companies to develop business plans for all new products. Elements of the business plan are:

1. The market analysis for the market segment targeted by the product, in terms of market development stage, competitive analysis and potential volume.
2. The marketing strategy in penetrating the target market, whether to compete on price, features, performance, or quality.
3. The development plans in terms of the chosen technology and architecture, people and equipment, tooling and material requirements.

THE TOTAL PRODUCT DEVELOPMENT PROCESS

4. The manufacturing plan of how and what is required to produce the product. The supplier's strategy, the fabrication and assembly processes and equipment, and the test and quality plans.
5. The product support plan in terms of field support and training, product repairs and warranty strategy, and impact on support for existing products.
6. The financial analysis and projected return of investment for the new product: the product development costs, the expected manufacturing and support costs, the warranty and service levels, as well as the economic impact on existing products.

In addition to the business plan, the development of a prototype or mock-up for the product is important to demonstrate the idea or the product feasibility to the management and give potential customers a change to comment on the utility of the product. There are various rapid prototyping techniques using physical mock-ups, software for screen generators, and command and transaction modelling. The advent of advanced three-dimensional mechanical computer-aided design (CAD) stations with rendering capabilities can produce a three-dimensional image of the product on a computer screen. In addition, the use of fourth-generation software languages and automatic screen generators and transaction simulators can speed up the use of software tools to simulate a mixed hardware/software product to potential users.

There are certain criteria for "go–no go" decision points to proceed into the development stage;

- Market requirements identified.
- Definition of the new product and its release schedule to meet market needs.
- Chosen technology and architecture are acceptable.
- Technical feasibility demonstrated through a working model or prototype.
- Planned levels of price, performance and reliability are acceptable.
- Adequate project return on investment (ROI).

These criteria have been discussed previously. Although there are no rules for the correct method of ensuring that a product is accepted into development, many factors predominate: the current financial status of the company, whether the management is willing to take a chance at this particular time, the credibility and previous track record of the project team and its leader, and the current competitive situation of the industry.

It is well understood that a certain percentage of the products do not go into development at this point, even if the product idea is sound, because of the company's current financial record or competitive situation and sometimes because of poor preparation by the project team. The company managers are looking for a particular ROI which is in line with the financial conditions in the industry but could tolerate a lower rate in the hope of landing a "star" product in the future. Unfortunately, less than 20 percent of new products end up as major successes in the electronics industry.

Once the go-ahead is given from concept to development, another iterative process begins: that of prototype and early production models documentation, building, testing, and verification. The number of iterations is very important to the overall schedule; some elements such as very large scale integration (VLSI) and plastics tooling can take a long time to fabricate, so that reducing the number of iterations has become an extremely important part of the design process. The current interest of DFM is due to the tendency to reduce the development phase of the design cycle.

The release to manufacturing is usually a highly formalized process of meeting specific internal, as well as external specifications, especially in government-regulated industries such as medical and defense industries. At the release checkpoint meeting, usually a long list of items is covered to ensure that the soon-to-be-released product meets all specifications, and internal and external requirements. A sample release to manufacturing items list is provided in Table 2.1.

The release point is a good measure of the project team's preparations and the company's approach to the design process. If the project development process is structured such that it can be divided into credible segments with identifiable goals, and the project team have faithfully carried out their assignments on schedule, then the meeting is a formality. Otherwise, the release to manufacturing would either be delayed or forced prematurely, resulting in serious revenue and customer satisfaction issues for the manufacturers.

2.3.1 Managing the Total Product Development Process

Managers are concerned about the development time, cost, and performance to specification of the electronics product development process, and with good reason. There is a well-documented history of electronics projects that were late, were more expensive than projected, and missed their deadlines for customer shipments. Many tools and techniques have been developed to ensure the success of new product development. The remainder of this chapter will be devoted to those tools and techniques.

Other industries have successfully solved this problem of making products on time, to schedule, and within specifications. The construction industry, as an example, can successfully complete skyscrapers, bridges, and plants, even within complex bidding processes and rules. That industry's secret is

TABLE 2.1. A summary of the Various Activities and the Areas of Responsibility

R & D	Product marketing	Manufacturing engineering	Product assurance	Technical marketing
Product meets all environmental specifications	Data sheet completed approved	Proper sourcing for all parts completed	Product meets all environmental specifications	Product support package ready
Disposition of unused stock	Pricing approved	Final parts stocking levels forecasted	Life test plans and preliminary results reviewed	Training begun—Sales and Service Engineers
System compatibility tests completed and signed off	Pricing complete	All manufacturing documentation complete and specifications	Product assurance/ materials engineering approval	All manuals repro ready
Post manufacturing shipment release review date set		All production tooling operational	All environmental tests complete	Support plan distributed to field
Internal maintenance specifications completed		Assembly/test procedures complete and in use	Regulatory agency approval	Final sales inspection/ test OK
Life test plans and preliminary results reviewed		Shipping packages ready		
All environmental tests complete		Incoming inspection procedure complete		
Product qualification test procedures performed and analyzed for design margins				

the division of the construction tasks into small, measurable elements, augmented by the use of proven techniques, and managed by experienced personnel through simple project tracking tools.

Figure 2.10 is a summary of typical scenarios that occur when a project is not on target. It shows the sensitivity of profits over the product life cycle, in terms of negative impact to profits from three possible causes: shipping 6 months later, mis-estimating production costs by 9% too low, and development cost overrun at 50%. Several assumptions are made: 20% annual growth rate, 12% annual price erosion, and a 5-year life of the product. Clearly, the most severe impact is due to the shipments being late, and the least is the cost overrun in product development.

The iterative nature of the product development cycle is mostly to blame for late delivery of products, since the "do the best you can" philosophy permeated the electronics industry in the early stages, partly because the engineering management then did not understand the complexities of the new technology on which the design engineers were working. In the 1970s, managers did not understand the impact of using integrated circuits and microprocessors on electronic designs, and in the 1980s managers could not fathom the complexities of using software in electronic products. They did not have personal experience with these technology-based innovations and therefore tended to minimize their engineers' concerns and reservations about project schedules. In addition, the engineers themselves had little experience in the new technologies, and were often too aggressive in their estimates.

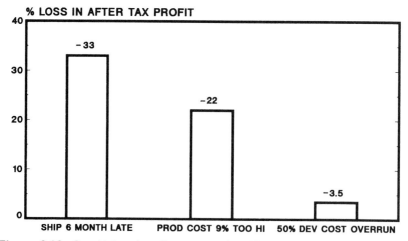

Figure 2.10. Sensitivity of profits over product life. (Assumptions: 20% growth rate; 12% annual price erosion; 5-year life.)

The concept of QFD (quality function deployment) is very useful in reducing the uncertainty of the iterative nature of the concept phase, and the tenets of concurrent engineering and design for manufacture are good methods for reducing the uncertainty of the development phase.

2.4 THE DESIGN PROJECT PHASES: MILESTONES AND CHECKPOINTS

The design phase is the most critical in terms of long-term profitability of new product development. Over the years, many companies have developed successful tools and techniques for design project management. The project tracking methodology will be discussed later in this chapter.

The design phase is usually broken down into small parts, with the completion of each part being recorded either in a formal checkpoint meeting, or at the completion of a particular task or milestone. It is important to have each checkpoint or milestone be of some significant and measurable progress in the project, to add to the project team's and the management's confidence in the progress towards achieving the product goals.

Historically, design project tasks have been divided along engineering disciplines: electrical, mechanical, software, and manufacturing. These disciplines were captured into distinct project groups, each with engineering responsibility over the discipline. This has sometimes led to interdisciplinary friction and factionalism. Another technique is the grouping of the engineers along project tasks or subparts, with interdisciplinary teams to encourage communication. Sometimes this is called *matrix management*.

The organization of the project team is very much dependent on the company's management philosophy, and some companies have found that both methods of organizing a project, along engineering disciplines or interdisciplinary teams, have been equally effective. It is important to manage these different activities positively by specifying the project interface among the different groups, and increasing the communication links by formal meetings, updates, and demonstration of phase and milestone completions.

These milestones for the project teams are an important tool used to update all project team members and the project management with team progress. The subdivision has taken the form of the following:

Breadboard: the product is demonstrated by a collection of working circuits, components, and mechanisms. It bears no physical resemblance to the final product, but can convey the functionality of the product, and perhaps can be used for software or interconnection system development.

Engineering model: the feasibility of the product and its functionality are demonstrated using some early production components, tooling, and assembly. Some functional and environmental testing could be performed on this product to measure compliance with specifications.

Pre-production prototype: the product and its components are procured and manufactured partly by final tooling and assembled by production personnel. This stage is used to finalize production documentation, methods, and techniques.

When there is a software component present in the product, the software engineers' efforts have to be integrated with the hardware engineers by having the early software version operate on one of the breadboard or engineering models. Thus the software development can proceed independently of the hardware effort, and the combined hardware/software system can be integrated fairly late in the product development. Conversely, the mechanical portion of the production has to be fairly well defined early in the cycle, because of the length of time necessary in developing the tooling and the mechanisms of the product.

The milestones for the project are the completion of tasks that meet internal and external requirements of the product. For example, internal requirements would be the completion of the different mock-ups, models, and working prototypes for customer and management approval. Additional requirements could be the testing of the product in a simulated environment to measure design robustness, prediction of future warranty performance, or ability to meet regulatory tests and specifications. Prototypes could also be used as demonstration units for the customers and the field service personnel, for feedback on customer use, regulatory evaluations, and field training and early sales development.

One of the important issues in ensuring success of a development project is the number of iterations of prototype building, analysis, and testing necessary to have the product meet its internal and advertised specification. Reducing the number of turnaround times is very important to keeping to schedule. Many companies have resorted to computer-aided engineering (CAE) workstations and software. In addition, the use of analysis software for verifying the design has been widely accepted recently. There are tools and techniques for reducing the iterations of design for each engineering discipline.

For electrical engineering, many of the iterations for printed circuit boards (PCBs) and integrated circuit (IC) designs have been due to some of the following factors:

- Lack of manufacturing knowledge in PCB and IC fabrication, assembly, and test. This could result in unacceptable rates of rejects in the manufacturing process.

THE DESIGN PROJECT PHASES: MILESTONES AND CHECKPOINTS 41

- The inexperience of the designers in making sure that the electronic design meets the components' specifications and their intended use.
- A poorly defined interface to outside and internal mechanisms, circuits, networks, and utilities.

Most of these problems can be resolved by the use of design guidelines for IC and PCB manufacturing, the use of simulation and analysis packages for analog and digital design, and the use of industry-wide standards for networking and interconnections.

In mechanical engineering, these iterations could be the result of similar causes: inadequate structural or thermal design considerations, or excessive vibration or mechanical wear requirements of the design. In addition, design iterations could be caused by specifying mechanical tolerances that are difficult or impossible to meet using the existing mechanical manufacturing equipment set, or by designing to improper tolerance stackup. The use of analysis packages such as finite elements, modal, and mold flow analysis can eliminate these potential problems.

In software, inadequate module specification and poor software design techniques can produce software that is full of errors or "bugs." If functionality is continually being added to the system, it will be difficult for the project team to complete the software effort. In some cases, poor development practices have resulted in software that needs to be continually updated in the field, done under the guise of "software enhancements."

It is important to recognize that using these tools of analysis and correct specification, sometimes called *verified design*, will lead to a longer initial design but a shorter overall design cycle. It is interesting that the amount of design time and analysis spent by the design engineer is only a small part of the overall design cycle. The majority of the time is spent on waiting for other support personnel and for building, testing, and evaluating the design. Use of the correct analysis tools in a verified design will lead to a shorter overall development time (Figure 2.11).

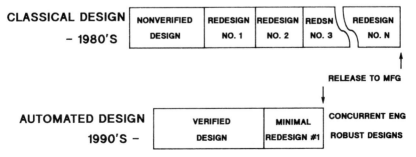

Figure 2.11. The impact of verified design cycles. Note the larger initial design cycle but shorter total development cycle.

2.5 PROJECT TRACKING AND CONTROL

The success of the product strategy will be measured ultimately through the profitability of the new product being developed. In addition to the expected sales volume and the duration of the product life cycle, two major cost elements determine the degree of success: the cost of the development project and the production cost of each unit. A formula can be developed to measure these values called the *return factor* (RF):

$$\text{RF} = \frac{\text{life cycle product operating profit}}{\text{product engineering development costs}} \quad (2.1)$$

Such formulas have historically been used to measure engineering investments for two major reasons: to ensure that the investments have an adequate rate of return in line with the traditional return for the business segment the company is operating within, and to select from competing projects owing to the finite financial resources of the company.

This formula is modified from the traditional formulas of measuring investments: the discounted cash flow method using present value (PV) on return of investment and a payback period:

$$\text{PV} = \sum_{i=1}^{n} \frac{\text{CF}_i}{(1+r)^i} \quad (2.2)$$

where
PV = present value
CF_i = cash flow for year i (product operating profit)
r = discount rate
n = number of years

It is important that the overall cost of the engineering development be included in the project costs. These are the salaries of the project team and the direct project charges and expenses such as tooling (software or hardware), model shop, materials purchase, and direct labor expended in building prototypes. Some of these costs could be recovered in the future since prototypes can be used for noncustomer sales such as field demonstration units or training beds for manufacturing or sales support. These data on project costs and return on investments should be summarized in a project data sheet. An example of a proposed project sheet is shown in Table 2.2, showing the progression of the project data sheet as the project is continuing under the different milestones, with the numbers evolving as the product gets close to release. Note that it is important to take a retrospective look 6 months to 1 year after release to production in order to examine how well the original estimates held up in the first months of production.

TABLE 2.2. Product Data Sheet

Product Name:		Product No:						
Project Manager:		Date:						
Project Staff: R & D	**Project Team Members** Manufacturing: Marketing: Quality: Field							
Milestone								Retr
Date (Estimated)								
Date (Actual)								
Development K$ Invested Estimated K$ To Go Total Development Costs								
Direct Project Charges Tooling Model Shop Material Direct labor								
Failure Rate (%/year) Warranty Cost (%/$ ship)								
Estimated Production Cost Estimated Selling Price Estimated Operating Profit								
Mature Sales (#/year) Mature VAS (K$/year) Impact on other products Net profit Expected (year)								
Return Factor (Profit/Cost—5 years)								

Capital equipment purchases such as engineering workstations, or general-use engineering support equipment such as CAD PCB layout systems, should not be included in the project cost base if they are to be used in future unrelated products. Similarly, general production automation equipment purchased for manufacturing productivity and quality should not be included in project cost factor calculation, but should be part of the company's capital expenditure plans. These capital costs are to be managed under the long-term financial plans of the company, and depreciated against the current operating statements.

The life cycle product operating profit should only include the incremental sales and profit due to the new product. Loss of existing product sales must be removed from the life cycle profit. In addition, life cycle costs of warranty and support costs must be included.

Several methods have been developed for ensuring that product development projects are on target to meet the project costs estimates, the manufacturing cost and quality targets, and achieve their project progress milestones. Most of these methods combine traditional project management techniques such as Pert and Gantt charts. These methods outline the duration for each project task and the earliest possible start. This start is usually contingent on completing a prerequisite task.

The project task data is then plotted versus time (Gantt chart) in linear fashion, with each project task start-point, duration, and end-point outlined versus time. The relationship between the different tasks, which is a hierarchy table showing which tasks are to be performed first and which ones cannot be started until others are completed, is not entirely visible to the chart user.

In a Pert chart plot, the task relationship is more symbolic and is not plotted against a time frame. There are project nodes, which may represent in symbolic form the start and finish of each task. Some of the tasks can form independent parallel paths between the project nodes. The one path in the project that takes the longest time is called the *critical path.* In project tasks not involved in the critical path, task slippage can occur up to the point where it would equal the time outlined by the critical path, without jeopardizing the project schedule.

Project tracking and control systems have to be updated frequently, to continuously examine whether the project is on target or is facing slippage. In the latter case, a new project chart has to be generated. Sometimes, project teams have resorted to regular monthly meetings to brief the entire project team and management to their progress in completing tasks and milestones. Each project group outlines its efforts in the preceding month and the tasks to be accomplished the following month. They extrapolate into the future any possible concerns and problems that could offset the project schedule.

PROJECT TRACKING AND CONTROL

Other tools of project tracking consist of computer-generated and updated data on project cost, milestones, and task completion. The most effective tools of project update are those that visually present the project status in one integrated chart, showing clearly the different points of progress. Two very effective graphical tools are the *bug* and *spider* (or *radar*) charts (Figures 2.12 and 2.13), so called because of their shapes.

The bug chart (Figure 2.12) is a plot of the project milestone versus the project expenses. The plot is arranged so that a project that is on track will plot as a straight line of 45 degrees. A project that is delayed in one of the milestone stages will show an offset, which will continue throughout the remainder of the milestones. The use of the project cost update as well as milestones is important not only for expense control but as a method of showing the timing of the acquisition of "big ticket" items that have been procured, such as tooling and supplier deliveries. The lack of progress in purchasing these items will seem to show up as a positive indication on project expenses. However, this could be very detrimental to the project schedule, as it results in a delay of accomplishing schedules tasks.

The spider (radar) chart (Figure 2.13) is a method of visual presentation of whether the project is meeting its separate goals in different areas through a one-year period. A project that is on track will have concentric circles at each time period, represented quarterly in the figure. The spider chart represents a quick visual check of all project goals and can be effective in spurring on different project groups to meet their goals concurrently with each other.

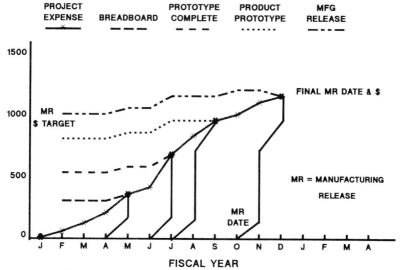

Figure 2.12. Project progress chart ("bug" chart).

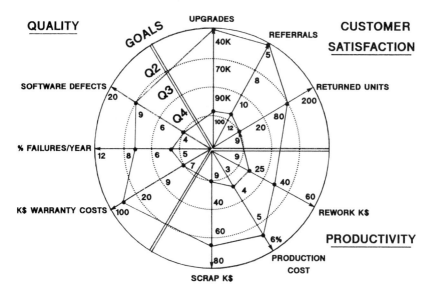

Figure 2.13. Spider (radar) chart.

2.6 CONCLUSION

This chapter has presented the total new product development cycle for electronic products in terms of the product life cycle, the impact of technology, and the development project tracking and control. The importance of the role of concurrent engineering and design for manufacturing should be clarified in the business plan and identified in specific project goals. The method of implementing the concurrent engineering and DFM in terms of methodology and tools will be reviewed in the upcoming chapters. It is of great impact to be specific about attaining the goals of concurrent engineering and manufacturability, in terms of setting design efficiency, product cost, reliability, and warranty targets for new products.

This chapter serves as the introduction of the product development stage, and the effects of technology and the marketplace on new products and their specifications. The future success for concurrent engineering can be achieved only in the context of efficiently concluding the overall product development plan. Understanding these issues of market strategy, technology impact, and project organization and control for the engineers is of great value in developing successful new products.

SUGGESTED READING

Andreasen, M. "Design Strategy." *Proceedings of the ICED*, Boston 1987. New York: ASME.
Cross, N. *Engineering Design Methods.* Somerset, N.J.: Wiley, 1989.
French, M.J. *Conceptual Design for Engineers.* London: Design Council Books.
Hays, Robert and Wheelwright, S. "Link manufacturing process and product lifecycles." *Harvard Business Review*, No. 79107, Jan.–Feb. 1979.
Kotler, Philip. *Marketing Management, Analysis Planning and Control.* Engelwood Cliffs, N.J.: Prentice-Hall, 1984.
Nevins and Whitney. *Concurrent Design of Products and Processes.* Hightstown, N.J.: McGraw-Hill, 1989.
Nichols, Keith. "Getting engineering changes under control." *Journal of Engineering Design*, Vol. 1, No. 1, 1990, pp. 5–15.
Ouchi, W.G. *Theory Z: How American Business Can Meet the Japanese Challenge.* Reading, Mass.: Addison Wesley, 1981.
Pugh, S. and Morley, I. *Towards a Theory of Total Design.* University of Strathclyde, Design Division, 1988.
Wheelwright, Steven C. "Operations as strategy: Lessons from Japan." *Stanford Graduate Business School*, Vol. 50, No. 1, Fall 1981/82.
Urban and Hauser, *Design and Manufacturing of New Products.* Engelwood Cliffs, N.J.: Prentice-Hall, 1985.

CHAPTER 3

Principles of Design for Manufacturing

The principles of design for manufacturing and design for electronic assembly are not new; childhood toys such as Tinker Toys, Lincoln Logs, and Lego are good examples of design for manufacturing as they have the same objectives:

Minimum parts types
Standard components
Parts fit/snap together
No fasteners
No assembly tools required
Reduced assembly time and operator skills

This chapter explores some of the principles and origins of design for manufacture (DFM) and will give examples of DFM guidelines. The axiomatic theory of design is used to highlight this methodology. These design guidelines were based on common lessons learned while developing electronic products. Many similar checklists are being used in major electronic companies as a repository for the collective wisdom of their successful design engineers.

These design guidelines emphasize the design of electronic products using self-locating and aligning parts built on a suitable base part. The number of parts should be minimized by using standard parts and integrating functionality and utility. Several features should be used, such as standard and automatic labeling, self-diagnosis capability at the lowest level, and use of symmetrical and tangle-free parts designs.

The IBM Proprinter, which when introduced in the mid-1980s was the state of the art in concurrent engineering and design for assembly (DFA), will be discussed as the prime example of a DFM product whose parts design

THE AXIOMATIC THEORY OF DESIGN 49

and process development is well integrated and whose assembly requires no tools or adjustments.

Finally several methods of setting and measuring the design for manufacturing process goals will be discussed. These setting processes and measures can be used to formally encourage the project teams to achieve quantifiable design for DFM/DFA improvements over the current product levels.

3.1 THE AXIOMATIC THEORY OF DESIGN

The axiomatic theory of design is a structured approach to implementing a product's design from a set of functional requirements. It was developed by Professor Nam Suh of MIT and outlined in his book, *The Principles of Design*. It is a mathematical approach to design which differentiates the attributes of successful products and demonstrates designs that are not manufacturable. The design process is seen as the development of functional requirements, which then can be mapped into design parameters through a matrix known as the *design matrix*, and then into manufacturing process variables. The functional requirements and the design parameters are hierarchical and should decompose into sub-requirements and sub-parameters.

The design function is bound by two types of constraints: input constraints, from the specifications of the product, and systems constraints, which are dictated by the environment where the product is used. These constraints act on the design parameters to form the boundary for the implementation of the design into physical parameters.

This approach is quite similar to the quality function deployment (QFD) approach, developed in Japan and popularized by the automobile industry. They share the structured concept of functional requirements, but QFD is implemented earlier in the product life cycle by focusing on customer inputs. QFD can be used as a guideline to designing evolutionary products, where the customer is already familiar with the product functionality, its intended use and the technology of design has reached a mature stage. Axiomatic design theory could be used in revolutionary products, where the design implementation is in the startup or growth stage.

There are only two axioms of design:

The Independence Axiom, which maintains the independence of functional requirements (FRs).
The Information Axiom, which minimizes the information content of the design. It is a mathematical version of the principle of KISS (known in the vernacular as "Keep it Simple, Stupid").

3.1.1 The Axiomatic Corollaries

The axiomatic corollaries are developed from the axiomatic theory of design and they bear strong resemblance to many design guidelines that were developed in manufacturing companies. The design guidelines that follow in the next section are grouped in the same categories as the axiomatic theory corollaries: decoupling of coupled designs, minimizing functional requirements, integration of physical parts, standardization, symmetry, and largest tolerance.

3.2 THE DESIGN GUIDELINES

The design guidelines presented below are a collection of guidelines specific for electronic products. These guidelines can be adopted in part or whole for different electronic products.

3.2.1 Decoupling of Coupled Designs

No Final Assembly Adjustments

Assembly of modules at final assembly with no adjustment to achieve system functions. This will reduce assembly cost, enhance automation, and facilitate maintenance and refurbishment.

Many electronic products are eliminating the need for final assembly adjustment. This is best done by a more robust design with controlled variability that does not require adjustments. Other methods include making the adjustment at the subassembly level, such as the alignment of cathode ray tubes (CRTs) prior to final assembly, or by having the customer perform the final assembly adjustments, such as that of the print head tension mechanism for the IBM Proprinter.

Self-Diagnosis Capability at Lowest Level

Provide capability to aid manufacturing in troubleshooting and self-diagnosis of machine problems, and to identify in the field parts/assemblies that have worn, drifted, or malfunctioned.

Examples of self-diagnosis are the use of microprocessors for self-diagnosis of the product each time it is turned on; the use of wear indicators, either mechanical or optical, to indicate time to replace worn parts; and the use of periodic monitoring of vital product subsystems to indicate problem areas. These techniques have been used widely in the electronics industry, with varying degrees of effectiveness.

THE DESIGN GUIDELINES 51

One of the important options for new product developers is the level of problem identification for self-diagnosis. There have been many attempts to isolate problems at the component level, most of which were not successful because the cost of extra circuitry and logic for component fault isolation far outweighs the benefits. Self-diagnosis at the printed circuit board (PCB) assembly level is the most common method, with individual PCBs having indicators such as light emitting diodes (LEDs) monitoring the PCBs functional status.

Most electronic self-diagnosis is activated by resident software on the PCB's microprocessor memory, or could be phoned in from a remote location, such as a regional or national service center. It is important that any diagnosis software for fault isolation be common for customers as well as the factory, to reduce the investment of new test software.

3.2.2 Minimizing Functional Requirements

Subsystem Modular Design

- A module is a discrete and self-contained assembly with a defined set of inputs on its performance.
- Components should be grouped as discrete modules used for similar products in the same family. Products share the same power supplies, fans, front panels, housings, and hardware.
- Subassemble for use in final assembly.
- Facilitate quality assessment prior to assembly.
- Multifunctional parts reduce assembly complexity.

Typically it is more efficient to use the minimum level of assemblies. The ideal product structure is the one where all components go into subassemblies, which then plug into final assembly. This structure eliminates the need to make interim scheduling of assembly levels and line fabrication items, and enhances the use of just-in-time (JIT) production.

A good partitioning of electronic products into functional and modular subassembly makes it easier to test the modules in process, and enhance the turn-on performance of the products at final assembly.

Build on Suitable Base and Stack Assembly

- Orient in horizontal plane.
- Provide stable base position and support.
- Place bottom part or base onto assembly fixturing.
- Establish part position with guides/aids: pins, recesses, insets, etc.

- Stack parts vertically. Assemble from above or divide into subassemblies. Take advantage of gravity.
- Avoid rotating or orienting or regripping parts.
- If fasteners must be used, apply and orient vertically.
- Use parts that are self-fixturing and aligning.
- Use pick-and-place operations.
- Allow for ease of insertion and removal, but consider serviceability.

The ideal electronic product does not require any tools or special gages to assemble. Parts should drop in, self-align, and snap together. It is important that as the assembly of the parts and subassemblies progresses, there is adequate direct access for production operators or service personnel for making adjustments and replacing or servicing subassemblies.

Minimize Level and Direction of Assembly

- Generate only those subassembly drawings that are required as either field spares or replacement by the manufacturing process. This results in reduced material handling and inventory control.
- Ideally assemble all parts in one direction: *down.*
- Unit should not be subjected to rotational and inertial loads.
- Avoid multi-motion insertion.
- Facilitate assembly by providing chamfers or tapers which will help to guide and position parts correctly.

The use of minimal directions of assembly facilitates the future automation of the assembly process, since the pick-and-place motions required are much simpler and could be performed by the less-complex robots. For manual operations, one-directional assembly facilitates tasks by allowing use of only one hand to manipulate each new part.

Using the minimum number of assembly levels and motions lessens the number of documents required to support the assembly process. When parts fit together using guides and chamfers, the need for assembly tools, diagrams, and instructions to operators is eliminated. This reduces the number of defects, as there is only one way to assemble the product adequately.

3.2.3 Integrating Physical Parts

Minimizing the Number of Parts

- Eliminate excess parts by applying three criteria:
    1. Must the part move relative to others parts in performing its function?
    2. Must the part be made of different material (strength, insulation, etc.)?

THE DESIGN GUIDELINES

3. Must the part be disassembled from the rest of the product (servicing, replacement, etc.)?

- Use value analysis tools to simplify designs:
  1. Identify necessary functions performed by each part.
  2. Find the most economical way to achieve function.
  3. Use checklist for parts design simplification.
- Brainstorm: Identify designs that require fewer parts; integrate functionality (use plastic, sintered materials, powder metallurgy, etc.)
- Choose the design for the best assembly method (base plus vertical stacking).

The number of parts is directly proportional to the cost of the product in terms of direct purchasing costs. Support costs such as purchasing and inventory of parts, supplier qualifications, warranty, and reliability estimates increase proportionality with the number of part numbers. Eliminating parts is one of the most important methods of reducing electronic product cost and enhancing quality.

The costs of new parts are usually not identified to the design engineer. They include the cost of setting up the part, selection of the suppliers, qualification of the part and the supplier, the cost of purchasing and keeping adequate inventory, the risks of obsolescence, and the reliability ratings. Every additional part contributes to the increase of mean time between failures (MBTF) of the product. It has been estimated that the setup of a new part could cost in excess of $5,000.

The three questions to be asked about the necessity for a new part were suggested by Boothroyd and Dewhurst in their book *Product Design For Assembly*. It is advisable to go through a checklist for new parts to encourage integration and functionality of new designs.

Examples of integrating parts are given in Figure 3.1. These examples show that extra parts could be eliminated using designs that would not only eliminate parts but also reduce assembly operation. As a result of these attempts at reducing parts, the suppliers and fabrication shops of electronics companies have been asked to make much more complex parts. The designers have to be careful when integrating new parts, because reducing assembly time can inversely affect the cost of fabricated parts.

Part design analysis checklist

What is the purpose of the part?
What is its location in assembly and its use?
Can another part be substituted for it?
Can it be combined with another part?

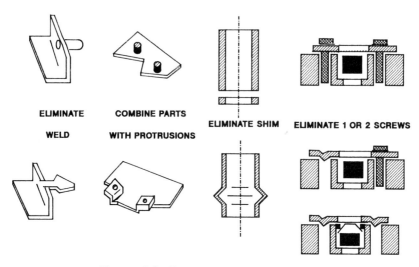

Figure 3.1. Examples of integrating parts.

Can the design do without it?
Can the material be changed?
Can it be made symmetrical?
Is it available at a lower cost (from supplier)?
Can it be quickly assembled or disassembled?
What does the competition do?
It is safe to handle (no sharp corners, edges)?
Is the assembly process compatible?
What other operation sequence can be used?
Are there interactions with other parts?
Is it solid for long use? Does it wear out?
Can it be repaired easily?
How does it work in the field environment?
Can it be grasped easily?
Can it be distinguished easily?
Does it have the same finish, tolerance, and material as the rest of the parts?

There are no rigorous guidelines for product design efficiency that can be applied across all products. The preceding checklist is intended to keep the parts designers focused on reducing the number of parts, and, if they are not able to do so, to integrate parts and make them more assembly-efficient and compatible with the current manufacturing process. The checklist is intended as a self-motivating guideline for development engineers to design more efficient parts and products.

THE DESIGN GUIDELINES

Design for lowest-cost fastening
- Whenever possible, use snap fits, press fits, slot/tabs, screws, and adhesives/welds, in that order.
- *Snap fits*

    Cons: tooling costs, tight tolerances, disassembly.
    Pros: Low cost; use CAE support packages for snap design.

- *Press fits.* Can lose force over time (creep). Requires tighter molding tolerance than snap fit. Lower tooling costs.
- *Slot/tab fits.* Use two interlocking parts such as slot and tab, or bent tab.
- *Screws.* If they must be used:

    Minimize numbers and types used.
    Use common head size, make room for drivers.
    Orient for vertical insertion.
    Avoid flat washers, integrate with lock washers.
    Use self-tapping screws.

- *Adhesives and welds.* These simplify design, but cause high assembly cost and are difficult to repair.

The elimination of screws and other fasteners is important in order to reduce the number of components and consequently the time of assembly and cost of new products. However, one important consequence is the increasing difficulty of disassembly of electronic products when snap fits are used. Sometimes this has led to use of special tooling for disassembly in order to reduce possible damage to the products. New product designers should carefully balance the two issues.

The use of adhesives and welds is appropriate when there is minimum space available for assembly or when hermetic sealing is required. The setting of parameters for adhesives or welds is important to achieving proper assembly. For adhesives, the parameters are the proper amount of adhesive to apply, the cleaning of parts before and after the adhesion operation, and the type of material to be used. For welding, the setting of the weld equipment in terms of electric current level, dwell time, and pressure is important. In both cases, the successful use of design of experiments has been helpful in optimizing the process.

Apply Labels Automatically
- Mold, paint, stamp, or engrave as part of assembly process. Avoid silk screen and detached metal labels.
- Use multi-purpose labels (combine information in one label).

- Program for serial markings and identification.
- Use bar codes or laser marking.
- Bar codes can be used for tracking and programming assembly.

The application of labels on the product is an important issue for registering the product through serial and model numbers for future warranty and repair history. Managing the labeling issue is critical, as label generation and application can be time-consuming in terms of new parts creation and its associated costs of documentation, purchasing, inventory, and storage. An additional benefit of applying automatic or bar code labels is that they can be used to set automatic routing and process parameters during the manufacturing of the product.

Cables

- *Avoid* cables because they are difficult to route and connect and may require hardware. Use prewired fans, direct connect, etc.
- Cables cause low reliability because they contain many parts, pins and wires.
- *If cables must be used*: Key the connectors, use positive latch mechanisms. Lock with snap feedback.
- Route away from sharp edges, bends, pinching, etc.
- Allow sufficient space to connect and to service unit.
- Use dummy plug to safeguard connector during assembly.

Cables are one of the hidden elements contributing to poor quality in new products. The number of components and assembly operations involved is large, leading to increased probability of defect creation. In many cases, final electronic testing of products will not cull out poor cable connections, which might eventually fail in the field. Many analyses of product defects in the field have shown that this low-technology problem is the cause of failures of many high-technology electronic products. Connectivity could be accomplished by using flexible PCBs or by mounting outside connectors (front and rear) directly into the PCBs.

3.2.4 Standardization

A standardization program of fasteners, components, materials, concentrics and finishes means:

  Commitment to a standard parts program. A goal should be that high percentage of parts used in a new design are *standard*. Standard parts do *not* mean obsolete parts.

  Reduction of setup of new parts. These costs can be substantial—as much as $5,000 per part.

Use modular and multifunctional parts

Group components in self-contained modules for similar products in same family.

Modular parts reduce assembly time and improve inventory costs.

Multifunctional parts combine several functions into one part at the assembly level and reduce complexity.

One of the main problems in controlling the number of standard parts is the lack of knowledge of designers of the existence of parts of identical or similar functions in the company's database. Although there are many part-numbering systems used to identify common types of components and parts, most are inadequate. For example, it is much easier for a designer to create, specify, and draw a new spacer bar than to go through many lists of existing product and part databases to discover whether a similar part is already used in the company.

The new concepts of *group technology* help designers to increase the use of standard parts. There are many computer-based systems that will retrieve the existing parts from the database and present CAD drawings of the parts to designers. These systems use specialized data compression and information query and retrieval techniques to reduce the amount of time and computer resources necessary to electronically identify and retrieve parts from the data base for design engineers. These systems could be stand-alone or integrated into the CAE/CAD/CAM architecture of the company. They are quite expensive, but at a $5,000 cost of implementing a new part through its life cycle, the rate of return on group technology systems making investing in them seem reasonable in order to reduce overall development costs.

One of the interesting measures of DFM is the setting of an aggressive goal for new designs by using a specified number such as 90% of existing and standard parts. This target seems unattainable, especially since technology is continuously providing updated electronic components. Upon close examination, many components of electronic products use standard basic parts such as resistors, capacitors, spacers, and existing IC logic family components. Therefore, the goal of using existing parts in 90% of each new design is tenable. The advantage of such a plan is to significantly lower new component setups and increase the quality of the new products. Established parts generate fewer defects, since their fabrication and assembly processes are well documented through previous use.

3.2.5 Symmetry

Use symmetrical parts, whenever possible

Easily handled parts promote assembly efficiency.

Symmetrical parts minimize handling.

Small changes can result in perfect symmetry.

Exaggerate asymmetrical parts

If parts are asymmetrical, they should have distinct polar properties by geometry or weight (think of the natural resting place for auto feeders) so that they can be separated, oriented, and fed easily, either manually or by automatic methods.

Avoid nesting and/or tangling of parts.

Change part designs to eliminate excess parts by using ribs, slots, close open spring ends, increasing dimensional thickness or angles, changing hole dimensions, etc.

It is easier for production operators and automatic assembly equipment to grasp symmetrical parts for presentation to the next assembly operation. Therefore, the use of symmetrical parts, whenever possible, or the addition of features to parts to make them single-axis, two-axis, or fully symmetrical, is recommended. In many cases, when the part is stamped out of sheet metal or molded by plastic injection, the extra cost of making the part symmetrical by adding extra holes or features is relatively low. The only adverse effect on the development cycle is the increased length of time it takes to add these features to the drawing and the subsequent machining of plastic molds. Examples are given in Figure 3.2.

Another element in the reduction of the grasping time is the elimination of nesting or tangling of parts. The addition of design features to reduce tangling and nesting, such as closing open spring ends and washers and

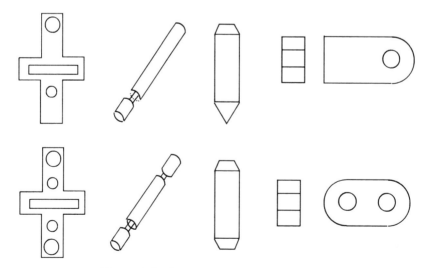

Figure 3.2. Examples of adding symmetry.

THE DESIGN GUIDELINES 59

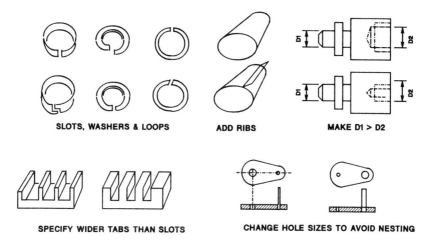

Figure 3.3. Avoidance of nesting or tangling.

altering some of the dimension and features of parts that cause nesting, can very easily be achieved in the design stage at a minimum cost to the development cycle, and can produce lower-cost assemblies for new products. Examples are given in Figure 3.3.

When symmetry cannot be achieved readily, or the cost of making the part symmetrical is too prohibitive, the asymmetry of the part should be exaggerated to make grasping of the part by manual or automatic means much easier. Examples of exaggerated asymmetrical designs are given in Figure 3.4.

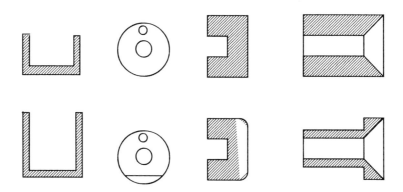

ASPECT RATIO   ADD FLAT   ADD RADIUS   CENTER OF GRAVITY

Figure 3.4. Increasing part asymmetry.

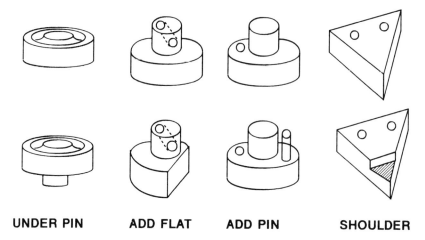

Figure 3.5. Increasing asymmetry to orient hidden features.

Another condition for using asymmetry is the exaggeration of a hidden feature that makes easier identification of the part orientation prior to grasping. Examples of such design features could be the addition of an underpin, or the addition of features that might identify a particular surface or position, such as a flat surface, a pin, or a shoulder added to a particular side. Examples of increasing asymmetry to orient hidden features are given in Figure 3.5.

3.2.6 Largest Tolerance
Self-Locating Features

Assembly of parts/modules to a precise location that eliminates the need for orientation or assembly adjustment. Examples include:

Tab in slots
Cutouts
D-shaped holes
Hard steps
Molded-in locaters
Guidepins
Dimples
Perforations
Design of a rigid part to act as "base" to which other parts are added to reduce tolerance stackup.

THE DESIGN GUIDELINES

The design tolerance of parts should include self-locating features with the widest possible tolerance to help in assembling the parts. The features mentioned above are some of the techniques that can be used to guide parts into final assembly locations, with one base part used to stack the other parts onto it.

Accommodate Errors, Maximize Compliance

Compliance, the accommodation of manufacturing error or uncertainty, should be designed into a product to avoid excessive assembly force. rework and scrap:

- Use only quality and therefore consistent parts.
- Specify adequate tolerances on all part drawings.
- Accommodate for mechanical wear.
- Use compliance techniques and guide features: chamfers, ramps, lead ins.

The relationship of the manufacturing tolerance to the part specification limits, called the *process capability index* is discussed in detail in Chapter 4. There should be a common goal for all engineers. Design engineers can increase the tolerance limits to the maximum that would still allow the proper functioning of the product, and manufacturing engineers should reduce the variability of the manufacturing process by using *statistical process control* (SPC) and other quality enhancement tools, by better equipment and tool replacement and maintenance schedules, and by using design of experiments to narrow the process variability. The ratio of their combined efforts is calculated in the process capability index.

Examples of self-locating and guiding features are given in Figure 3.6. These involve simple design changes that can make the assembly of parts much easier, especially by eliminating the use of special tools or techniques.

Limit Flexible Items

- Design parts with safety margins sufficient to prevent bending or collapse under the force of assembly.
- Wires, cables, and belts are difficult to handle, especially in automated assembly.
- Components should plug together, where possible.
- Use printed circuit boards to advantage in reducing flexible items.
- Use gears and shafts to substitute for belt drives.

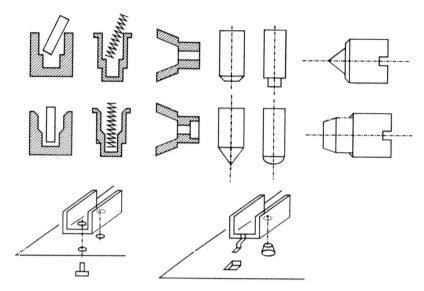

Figure 3.6. Self-locating and guiding features.

Flexible items are difficult to grasp, especially using automatic feeders, and therefore should be replaced by different designs whenever possible. Examples are the use of drive shafts instead of belts and the use of flexible spacer materials instead of soft grommets to fit a large-tolerance dimension.

3.3 A DFM EXAMPLE: THE IBM PROPRINTER

The IBM Proprinter is considered a prime example of an electronic product that was designed for flexible automated assembly. By focusing on concurrent engineering and design for manufacturing, IBM has been able to use manufacturing as a strategic weapon to achieve and maintain a market share in the highly competitive personal printer market, once dominated by Far East companies.

The early philosophy of the Proprinter was developed in 1982, when a high-level task force of five engineers and managers was formed at IBM to develop the product and the process manufacturing strategy. The design engineering effort consisted of 60 engineers divided into 20 subsystem design teams, each focused on individual subassemblies. Each team was made up with a design engineer, a manufacturing engineer, a moderator, and a technology expert, as needed.

A DFM EXAMPLE: THE IBM PROPRINTER

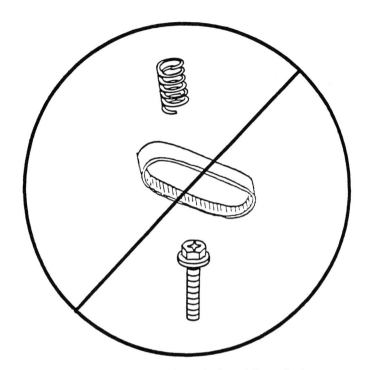

Figure 3.7. IBM Proprinter design philosophy logo.

The design goals were very simple, and the design team developed a logo for the project (see Figure 3.7). The criteria for the design were very simple:

Develop a modular design.
Minimize the total part count.
All parts to be self aligning for ease of assembly.

The design review cycle was updated to include mandatory DFM design reviews. Each DFM review took 4–8 hours to complete. A total of 50 design reviews were conducted, and the project took 6 months to complete.

Because of the DFM focus and review process, the design process took longer to be accomplished in the initial phases, but the project benefited in the long run from the increased manufacturability efforts. The manufacturing process was designed concurrently and implemented in a plant built specifically for the IBM Proprinter.

The IBM Proprinter represented the state of the art in design for assembly, although the goal of full downward assembly was not totally accomplished. The number of parts was far lower than for competing printers (Figure 3.8).

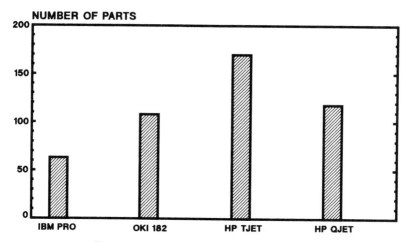

Figure 3.8. Personal printer parts counts.

Other principles of DFM were successfully implemented, including:

No fasteners.
No springs.
No pulleys.
No adjustments.
No excess modules.
No cables or harnesses.
No paint.
No labels.
No extra parts.
No multidirectional assembly.
No alignment/location tools or fixtures.
No external tests.
No engineering change notices (ECNs).
No custom builds or options.
No manual assembly.

Several major companies were so impressed by the Proprinter effort that they set about understanding and informing their engineers about the accomplishment and potential benefits of DFM to their products. These included Xerox Corporation, which prepared a report by their Automation Institute on the Proprinter design philosophy, and analyzed each subassembly. The Westinghouse Corporation Center for Quality and Productivity prepared a videotape, which was made available to Westinghouse engineers and to other companies, outlining the benefits of DFM and showing the manual assembly of a Proprinter in about 3 minutes, without the use of any tools or fixtures. The Proprinter remains today as the standard for DFM/DFA and concurrent engineering.

3.4 SETTING AND MEASURING THE DESIGN PROCESS GOALS

DFM goals for new product development usually take a historical progression in companies that are beginning to consider manufacturability and concurrent engineering as part of their competitive strategy.

The first attempt at DFM is the initiation of manufacturing, service, and reliability plans that encourage manufacturing's and other departments' feedback on new designs. These initial DFM goals are general in nature, and represent a commitment by the project team and their management to consider DFM as part of the design process. The DFM goals are not specific, and are usually "best effort" attempts to reduce manufacturing cost. The review for manufacturability becomes part of the design review process, as well as the review for testability, serviceability, and reliability (these will be referred to as "-ilities". There is a tendency to set manufacturability goals similar to the guidelines shown in this chapter.

Formal DFM goals should be set after the company has attempted some projects with DFM goals and has experienced initial successes. These can be specific manufacturing targets that are measurable and are in line with the other goals of the new product. However, the current levels of "-ilities" should be made available, either through existing databases and metrics or through actual measurement of current products. These current levels should become the baseline from which achievable new product DFM goals can be targeted. Some of these specific goals might be in the following form:

Set minimum design efficiency values, based on design efficiency methodologies to be discussed in Chapter 7. This percentage will be founded on baseline levels obtained from analysis of current designs.

Set the percentage of standard parts used in new products. The current percentage should be calculated from existing products, and the new product percentage should be set appropriately. A target of 90% is achievable for new electronic products.

Set the product specification limits, based on a target process capability index (see Chapter 4) for new product designs. The process capability index is a good methodology for getting both design and manufacturing engineering to achieve a common goal. A 100 parts per million quality level is achievable from a process capability index (C_p) of 1.3, and a six-sigma level of 2 parts per billion is achievable from a C_p of 2.

Set the target number of suppliers, part numbers, and total parts, based on competitive analysis and current product levels. The reduction of the number of suppliers encourages closer relationship with the remaining suppliers, with the aim of improving quality, delivery, and material costs. The reduction of part numbers reduces the cost of setting up new parts and inventory costs and increases product reliability. A typical scheme reduces the number of suppliers by a factor of 2 to 3.

Set the final product assembly and test times, at a level that is lower than for comparable current products. This will encourage the design engineers to use parts that are easy to automate and assemble. Examples of these are modular assemblies, minimum part counts, and symmetrical, easy-to-align parts. Set first-pass yields at PCB automatic test (including functional tests) to 95%, and the level of product first turn-on at final assembly to 90%.

Set the warranty costs of new products, normalized by either defects/K$ or defects per product, at a level significantly lower than for current products. Set the mean time between failures (MTBF) similarly. The quality goals of the company should be in line with the competition and their progress. Increasing the quality level by a factor of 50% for every product cycle (3 to 5 years) is aggressive but achievable for the electronics industry.

These goals have the advantage of being measurable at different stages of the development cycle. Any combination of these goals will encourage the design team to use the design guidelines outlined in this chapter. Progress towards these measurable goals should be reviewed at the different milestones of the project. Formalizing the DFM principles in the project development phase will ensure that new products will be designed for manufacture.

3.5 CONCLUSION

This chapter is intended to act as a model for developing design guidelines for manufacturing and field performance of electronic products. These goals should be institutionalized in terms of setting very specific and measurable goals, rather than a "best effort" attempt at implementing these guidelines. The performance of existing products in terms of these goals should be quantified, and act as baseline for setting new product goals. The new product development process should include design reviews for manufacturability and field performance. At that time, progress towards these goals should be quantified.

New product development project teams should focus on these strategies by setting measurable and attainable goals for the product performance in factory and field. They should be set using current products' performance as the baseline with aggressive but achievable targets.

REFERENCES AND SUGGESTED READING

Andreasen, M. and Hein, L. *Integrated Product Development.* Kempston: IFS Publications Ltd., 1989.

Andreasen, M., Kahler, S. and Lund, T. *Design for Assembly.* Kempston: IFS Publications Ltd., 1983.

Botticelli, A. "A design guide to precision sheet metal fabrication." *Electronic Packaging and Production Magazine,* July, 1980.

Boothroyd, G. and Dewhurst, P. *Product Design for Assembly.* Wakefield, R.I., 1987.

Dangerfield, K., "The Management of Design." *Proceedings of the 6th Annual Conference on Design Engineering.* Kempston: IFS Publications Ltd., 1988.

Eder, W.E. "Theory of technical systems: Prerequisite to design theory." *Proceedings of the Institution of Mechanical Engineers.* Boston: ICED, 1987.

Holden, Happy. "DFM, the competitive program for the 90s." Lecture given at the INTEREX Conference, Boston, Mass., August 1990.

Miles, B.L. "Design for assembly: A key element within the design for manufacture." *Proceedings of the Institution for Mechanical Engineers,* 203, 1988.

Suh, Nam P. *The Principles of Design.* New York: Oxford University Press, 1990.

CHAPTER 4

Product Specifications and Manufacturing Process Tolerances

The process capability index C_p is gaining acceptance as a measure of product design for manufacturing, especially in the process-intensive industries such as IC and PCB fabrication. Historically, it has been one of the earliest methods of measuring design for manufacture by forecasting the reject rates in production of new components, assemblies, and products. Its unique blend of the production variability versus design specification makes it a natural method for setting, communicating, and comparing new product specifications and manufacturing quality levels for competitive manufacturing plants.

By focusing on the process capability index, there is a commitment up-front to measuring and controlling manufacturing variability through statistical process control (SPC) tools and methods such as control charts. In addition, it is an excellent tool for negotiating and communication with suppliers to set the appropriate quality level and expectations.

The process capability index focuses on communication between the design, development, and manufacturing parts of the organization. By managing the relationship of design tolerance to manufacturing specifications, it shifts attention away from a possible adversarial relationship between design and manufacturing to a more constructive one, where the common goal of achieving a particular index level facilitates negotiations and cooperation in new product development.

4.1 THE DEFINITIONS OF TOLERANCE LIMITS AND PROCESS CAPABILITY

Electronic products are manufactured through materials and processes that are inherently variable. Design engineers specify materials and process

THE DEFINITIONS OF TOLERANCE LIMITS AND PROCESS CAPABILITY

characteristics to a *nominal* value, which is the ideal level for use in the product. The maximum range of variation of the product characteristic that will still work in the product determines the tolerance about that nominal value. This range is expressed as upper and lower specifications limits (USL and LSL)—see Figure 4.1.

The manufacturing process variability is usually approximated by a normal probability distribution, with mean of U and a standard deviation of σ. The process capability is defined as the full range of normal manufacturing process variation measured for a chosen characteristic. Assuming normal distribution, 99.74% of the process output lies between -3σ and $+3\sigma$.

A properly controlled manufacturing process should make products whose output mean characteristic or target are set to the nominal value of the specification. This is easily achieved through control charts. If the process mean is not equal to the product nominal value, it can be shifted by recalibrating production machinery or inspecting incoming raw material, characteristics.

The variation of the manufacturing processes (process capability) should be well within the product tolerance limits. The intersection of the process capability and the specification limits determines the reject level (Figure 4.2). Process capability can be monitored using control charts. The manufacturing process variability can be reduced by using optimized equipment calibration and maintenance schedules, increased material inspection and testing, and by using design of experiments to determine the best set of process parameters to reduce variability.

The classical design for manufacturing conflict of interests between design and manufacturing engineers is usually about controlling product quality and cost. The design engineers would prefer the narrowest possible process capability, so that they can specify the maximum tolerance specifications to ensure the proper functioning of their product. In contrast, the manufacturing engineers would prefer the widest possible tolerance specification, so that they can continue to operate the largest possible manufacturing capability

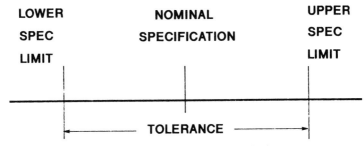

Figure 4.1. Process capability index.

70 PRODUCT SPECIFICATIONS AND MANUFACTURING PROCESS TOLERANCES

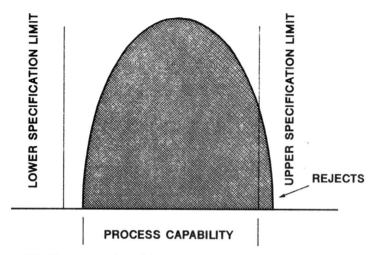

Figure 4.2. The intersection of the process capability and the specification limits to determine the reject level.

to reduce the amount of rejects. The process capability index is a good arbiter of the two groups' interests.

4.2 THE RELATIONSHIP BETWEEN MANUFACTURING VARIABILITY AND PRODUCT SPECIFICATIONS FOR NEW PRODUCTS

The ideal manufacturing process should produce clones of the production by performing replication of all fabrication and assembly materials, processes, and movements. However, this can never be achieved, because of variations in manufacturing: As the production machines and processes continue to turn out the product, the characteristics of materials and tools in the process change as they are being consumed. Materials hardness can change from the supplier for different lots and still be within specifications; machinery, fixtures, and tools wear out; and even though they can be replaced, recalibrated, or resharpened in an ideal maintenance schedule, they can still result in variation in the product. Employees can be properly trained to perform production tasks, but will slightly alter production operations because of fatigue or human error. Conditions beyond the control of the plant management could result in variability, due to environmental and weather changes or changes in suppliers, which are further multiplied by their subsuppliers' variations.

Some or all of these conditions can cause product variability, which when added up at each level of production could cause some of the product to become defective even though it is within acceptable limits at each stage of production. This reject rate will adversely effect the quality, and hence the cost, of the product.

There are two ways to increase the quality level of new products: Either increase the product specification limit and allow manufacturing variability to remain the same yet produce fewer defects; or reduce manufacturing variability by improving the quality level of materials and processes through inspection, increased maintenance, and performing design of experiments to determine variability sources and counteract them. The ratio of the interaction of these two sources of rejects is called the process capability index, C_p:

$$C_p = \frac{\text{specification width (or design tolerance)}}{\text{process capability (or total process variation)}}$$

$$C_p = \frac{\text{USL} - \text{LSL}}{6\sigma(\text{total range from } -3\sigma \text{ to } +3\sigma)}$$

(4.1)

where
USL = upper specification limit
LSL = lower specification limit
σ = manufacturing process standard deviation

The C_p value can predict the reject rate of new products by using normal probability distribution curves. A high C_p index indicates that the process is capable of replicating faithfully the product characteristics, and therefore will produce products with high quality.

The utility of the C_p index is that it shows the balance of the quality responsibility between the design and manufacturing engineers. The quality level is set by the ratio of the effects of both. The design engineer should increase the allowable tolerance to the maximum value that still permits the successful functioning of the product, and the manufacturing engineer should minimize the variability of the manufacturing process by proper material and process selection, inspection, calibration, and control, and by performing design of experiments.

An example in the electronics industry is the physical implementation of electronic designs in printed circuit board (PCB) layout. The design engineer might select a higher number of layers in a multilayer PCB, which will speed up the layout process because each additional layer increases the availability of making electrical connections. Minimizing the number of layers could possibly result in more layout time, but would produce lower-cost PCBs and fewer rejects, because there are fewer process steps. This is a classical case of focusing on project development expediency to the detriment of manufacturing cost and quality.

4.2.1 Determining the Predicted Reject Level from the C_p Index

The manufacturing reject rate can be determined from elements used in the definition of the process capability index C_p. By stating the design tolerance specifications in terms of process capability (or σ, the manufacturing standard deviation), it is possible to determine manufacturing reject rate from the normal probability distribution tables (Table 4.1).

TABLE 4.1. Cumulative Distribution Function for the Standard Normal Distribution (SND)

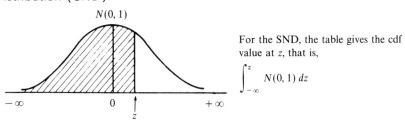

For the SND, the table gives the cdf value at z, that is,

$$\int_{-\infty}^{z} N(0,1)\, dz$$

z	0.09	0.08	0.07	0.06	0.05	0.04	0.03	0.02	0.01	0.00
−3.5	0.00017	0.00017	0.00018	0.00019	0.00019	0.00020	0.00021	0.00022	0.00022	0.00023
−3.4	0.00024	0.00025	0.00026	0.00027	0.00028	0.00029	0.00030	0.00031	0.00033	0.00034
−3.3	0.00035	0.00036	0.00038	0.00039	0.00040	0.00042	0.00043	0.00045	0.00047	0.00048
−3.2	0.00050	0.00052	0.00054	0.00056	0.00058	0.00060	0.00062	0.00064	0.00066	0.00069
−3.1	0.00071	0.00074	0.00076	0.00079	0.00082	0.00085	0.00087	0.00090	0.00094	0.00097
−3.0	0.00100	0.00104	0.00107	0.00111	0.00114	0.00118	0.00122	0.00126	0.00131	0.00135
−2.9	0.0014	0.0014	0.0015	0.0015	0.0016	0.0016	0.0017	0.0017	0.0018	0.0019
−2.8	0.0019	0.0020	0.0021	0.0021	0.0022	0.0023	0.0023	0.0024	0.0025	0.0026
−2.7	0.0026	0.0027	0.0028	0.0029	0.0030	0.0031	0.0032	0.0033	0.0034	0.0035
−2.6	0.0036	0.0037	0.0038	0.0039	0.0040	0.0041	0.0043	0.0044	0.0045	0.0047
−2.5	0.0048	0.0049	0.0051	0.0052	0.0054	0.0055	0.0057	0.0059	0.0060	0.0062
−2.4	0.0064	0.0066	0.0068	0.0069	0.0071	0.0073	0.0075	0.0078	0.0080	0.0082
−2.3	0.0084	0.0087	0.0089	0.0091	0.0094	0.0096	0.0099	0.0102	0.0104	0.0107
−2.2	0.0110	0.0113	0.0116	0.0119	0.0122	0.0125	0.0129	0.0132	0.0136	0.0139
−2.1	0.0143	0.0146	0.0150	0.0154	0.0158	0.0162	0.0166	0.0170	0.0174	0.0179
−2.0	0.0183	0.0188	0.0192	0.0197	0.0202	0.0207	0.0212	0.0217	0.0222	0.0228
−1.9	0.0233	0.0239	0.0244	0.0250	0.0256	0.0262	0.0268	0.0274	0.0281	0.0287
−1.8	0.0294	0.0301	0.0307	0.0314	0.0322	0.0329	0.0336	0.0344	0.0351	0.0359
−1.7	0.0367	0.0375	0.0384	0.0392	0.0401	0.0409	0.0418	0.0427	0.0436	0.0446
−1.6	0.0455	0.0465	0.0475	0.0485	0.0495	0.0505	0.0516	0.0526	0.0537	0.0548
−1.5	0.0559	0.0571	0.0582	0.0594	0.0606	0.0618	0.0630	0.0643	0.0655	0.0668
−1.4	0.0681	0.0694	0.0708	0.0721	0.0735	0.0749	0.0764	0.0778	0.0793	0.0808
−1.3	0.0823	0.0838	0.0853	0.0869	0.0885	0.0901	0.0918	0.0934	0.0951	0.0968
−1.2	0.0985	0.1003	0.1020	0.1038	0.1057	0.1075	0.1093	0.1112	0.1131	0.1151
−1.1	0.1170	0.1190	0.1210	0.1230	0.1251	0.1271	0.1292	0.1314	0.1335	0.1357
−1.0	0.1379	0.1401	0.1423	0.1446	0.1469	0.1492	0.1515	0.1539	0.1562	0.1587
−0.9	0.1611	0.1635	0.1660	0.1685	0.1711	0.1736	0.1762	0.1788	0.1814	0.1841
−0.8	0.1867	0.1894	0.1922	0.1949	0.1977	0.2005	0.2033	0.2061	0.2090	0.2119
−0.7	0.2148	0.2177	0.2207	0.2236	0.2266	0.2297	0.2327	0.2358	0.2389	0.2420
−0.6	0.2451	0.2483	0.2514	0.2546	0.2578	0.2611	0.2643	0.2676	0.2709	0.2743

MANUFACTURING VARIABILITY AND PRODUCT SPECIFICATIONS

z	0.00	0.01	0.02	0.03	0.04	0.05	0.06	0.07	0.08	0.09
−0.5	0.2776	0.2810	0.2843	0.2877	0.2912	0.2946	0.2981	0.3015	0.3030	0.3085
−0.4	0.3121	0.3156	0.3192	0.3228	0.3264	0.3300	0.3336	0.3372	0.3409	0.3446
−0.3	0.3483	0.3520	0.3557	0.3594	0.3632	0.3669	0.3707	0.3745	0.3783	0.3821
−0.2	0.3859	0.3897	0.3936	0.3974	0.4013	0.4052	0.4090	0.4129	0.4168	0.4207
−0.1	0.4247	0.4286	0.4325	0.4364	0.4404	0.4443	0.4483	0.4522	0.4562	0.4602
−0.0	0.4641	0.4681	0.4721	0.4761	0.4801	0.4840	0.4880	0.4920	0.4960	0.5000

z	0.00	0.01	0.02	0.03	0.04	0.05	0.06	0.07	0.08	0.09
+0.0	0.5000	0.5040	0.5080	0.5120	0.5160	0.5199	0.5239	0.5279	0.5319	0.5359
+0.1	0.5398	0.5438	0.5478	0.5517	0.5557	0.5596	0.5636	0.5675	0.5714	0.5753
+0.2	0.5793	0.5832	0.5871	0.5910	0.5948	0.5987	0.6026	0.6064	0.6103	0.6141
+0.3	0.6179	0.6217	0.6255	0.6293	0.6331	0.6368	0.6406	0.6443	0.6480	0.6517
+0.4	0.6554	0.6591	0.6628	0.6664	0.6700	0.6736	0.6772	0.6808	0.6844	0.6879
+0.5	0.6915	0.6950	0.6985	0.7019	0.7054	0.7088	0.7123	0.7157	0.7190	0.7224
+0.6	0.7257	0.7291	0.7324	0.7357	0.7389	0.7422	0.7454	0.7486	0.7517	0.7549
+0.7	0.7580	0.7611	0.7642	0.7673	0.7704	0.7734	0.7764	0.7794	0.7823	0.7852
+0.8	0.7881	0.7910	0.7939	0.7967	0.7995	0.8023	0.8051	0.8079	0.8106	0.8133
+0.9	0.8159	0.8186	0.8212	0.8238	0.8264	0.8289	0.8315	0.8340	0.8365	0.8389
+1.0	0.8413	0.8438	0.8461	0.8485	0.8508	0.8531	0.8554	0.8577	0.8599	0.8621
+1.1	0.8643	0.8665	0.8686	0.8708	0.8729	0.8749	0.8770	0.8790	0.8810	0.8830
+1.2	0.8849	0.8869	0.8888	0.8907	0.8925	0.8944	0.8962	0.8980	0.8997	0.9015
+1.3	0.9032	0.9049	0.9066	0.9082	0.9099	0.9115	0.9131	0.9147	0.9162	0.9177
+1.4	0.9192	0.9207	0.9222	0.9236	0.9251	0.9265	0.9279	0.9292	0.9306	0.9319
+1.5	0.9332	0.9345	0.9357	0.9370	0.9382	0.9394	0.9406	0.9418	0.9429	0.9441
+1.6	0.9452	0.9463	0.9474	0.9484	0.9495	0.9505	0.9515	0.9525	0.9535	0.9545
+1.7	0.9554	0.9564	0.9573	0.9582	0.9591	0.9599	0.9608	9.9616	0.9625	0.9633
+1.8	0.9641	0.9649	0.9656	0.9664	0.9671	0.9678	0.9686	0.9693	0.9699	0.9706
+1.9	0.9713	0.9719	0.9726	0.9732	0.9738	0.9744	0.9750	0.9756	0.9761	0.9767
+2.0	0.9773	0.9778	0.9783	0.9788	0.9793	0.9798	0.9803	0.9808	0.9812	0.9817
+2.1	0.9821	0.9826	0.9830	0.9834	0.9838	0.9842	0.9846	0.9850	0.9854	0.9857
+2.2	0.9861	0.9864	0.9868	0.9871	0.9875	0.9878	0.9881	0.9884	0.9887	0.9890
+2.3	0.9893	0.9896	0.9898	0.9901	0.9904	0.9906	0.9909	0.9911	0.9913	0.9916
+2.4	0.9918	0.9920	0.9922	0.9925	0.9927	0.9929	0.9931	0.9932	0.9934	0.9936
+2.5	0.9938	0.9940	0.9941	0.9943	0.9945	0.9946	0.9948	0.9949	0.9951	0.9952
+2.6	0.9953	0.9955	0.9956	0.9957	0.9959	0.9960	0.9961	0.9962	0.9963	0.9964
+2.7	0.9965	0.9966	0.9967	0.9968	0.9969	0.9970	0.9971	0.9972	0.9973	0.9974
+2.8	0.9974	0.9975	0.9976	0.9977	0.9977	0.9978	0.9979	0.9979	0.9980	0.9981
+2.9	0.9981	0.9982	0.9983	0.9983	0.9984	0.9984	0.9985	0.9985	0.9986	0.9986
+3.0	0.99865	0.99869	0.99874	0.99878	0.99882	0.99886	0.99889	0.99893	0.99896	0.99900
+3.1	0.99903	0.99906	0.99910	0.99913	0.99915	0.99918	0.99921	0.99924	0.99926	0.99929
+3.2	0.99931	0.99934	0.99936	0.99938	0.99940	0.99942	0.99944	0.99946	0.99948	0.99950
+3.3	0.99952	0.99953	0.99955	0.99957	0.99958	0.99960	0.99961	0.99962	0.99964	0.99965
+3.4	0.99966	0.99967	0.99969	0.99970	0.99971	0.99972	0.99973	0.99974	0.99975	0.99976
+3.5	0.99977	0.99978	0.99978	0.99979	0.99980	0.99981	0.99981	0.99982	0.99983	0.99983
+3.6	0.9998409									
+3.7	0.9998922									
+3.8	0.9999276									
+3.9	0.9999519									
+4.0	0.9999683									
+4.1	0.9999793									
+4.2	0.9999867									
+4.3	0.9999915									
+4.4	0.9999966									
+5.0	0.9999997									
+6.0	0.999999999									

The normal tables are set in terms of the area under the curve at a particular point where the probability that the value of the parameter is higher than a specified X coordinate value. This probability area is determined by the mean U and the standard deviation σ. This is called the Z-based distribution:

$$Z = \frac{X - U}{\sigma}$$

Examples of Using Probability Distributions to Calculate Defect Rates

The following example shows the calculations for the reject rate based on a C_p index of 1. This C_p value sets the tolerance specification limits to $\pm 3\sigma$. Using the normal distribution tables of Z value, with X being 3σ and U equal to 0 (the manufacturing process mean equal to the nominal value), the probability of a parameter value being greater than the upper specification limit (USL), and therefore a reject, is equal to

$$Z = \frac{3\sigma - 0}{\sigma} = 3 \tag{4.2}$$

$P(Z) = P(3) = 0.00135$ or 0.135% probability, from the probability tables of area under the probability curve at $Z = 3$.

Since the manufacturing probability distribution is assumed centered, the probability of a parameter being less than the lower specification limit (LSL) is equal to the probability of USL, for a total reject percentage of 0.26%. Translating this percentage to the more common part per million or ppm, it is equal to 0.26×10^4 or 2,600 ppm.

Conversely, aiming at a 100 ppm reject rate for a new product, the tolerance specification limits can be obtained by the following method, assuming that the process is centered around the nominal value of the specifications.

$$\begin{aligned} &\text{Reject rate at the USL} = 100\,\text{ppm}/2 = 50\,\text{ppm} \\ &P(Z) = 1 - (50/1{,}000{,}000) = 1 - 0.000050 = 0.99995 \\ &Z = 3.9 \quad \text{or} \quad \text{tolerance limit} = \pm 3.9\sigma \\ &C_p = \frac{\pm 3.9\sigma}{\pm 3\sigma} = 1.3 \end{aligned} \tag{4.3}$$

Note that in all of these calculations the manufacturing process is assumed centered and equal to the nominal value. Therefore, the reject rate of product characteristic above the USL is equal to the reject rate below the LSL. If the process is not centered, then the C_p index is assumed to be the higher of the two indices generated by taking the upper or lower specification limits

MANUFACTURING VARIABILITY AND PRODUCT SPECIFICATIONS 75

and subtracting them from the process mean U:

$$C_p = \frac{\text{USL} - U}{3\sigma} \tag{4.4}$$

or

$$C_p = \frac{U - \text{LSL}}{3\sigma} \tag{4.5}$$

Yet another index, called C_{pk}, determines the process capability based on the shift of the manufacturing process mean to the nominal value of the specifications. The relationship of the two indices is as follows:

$$C_{pk} = C_p(1 - K) \tag{4.6}$$

where

$$K = \frac{\text{nominal} - U}{(\text{USL} - \text{LSL})/2}$$

In Table 4.2 the part design specification as well as the manufacturing process characteristics of mean and standard deviation (StdD) are given as inputs. Using the formulas and examples in this part of the chapter, the defect parameters can be calculated as follows:

Start by calculating the defects %above and %below the specification. In each case use the formula for $P(\mathbf{Z})$, with Z being formed by:

$$Z = \frac{X - U}{\sigma} \tag{4.7}$$

TABLE 4.2. Examples of Calculation of Defect Rates

Specification nominal ± tolerance	Process Mean	StdD	%OK	% above USL	% below LSL	C_p or C_{pk}
10.00 ± 0.02	10.00	0.01	95.44	2.28	2.28	0.667
10.00 ± 0.02	9.99	0.01	84.0	0.13	15.87	0.667
10.00 ± 0.03	10.00	0.01	99.74	0.13	0.13	1.0
15.00 ± 0.05	15.00	0.02	98.76	0.62	0.62	0.83
8.00 ± 0.01	8.00	0.04	19.74	40.13	40.13	0.083
10.00 ± 0.02	10.01	0.01	84.0	15.87	0.13	0.667
10.00 ± 0.03	10.02	0.01	84.1	15.87	0	1.0

76 PRODUCT SPECIFICATIONS AND MANUFACTURING PROCESS TOLERANCES

Use X as the specification mean + upper or lower limit
Use U as the manufacturing process mean
Use σ as the manufacturing process standard deviation.

$$\% \text{ above or } \% \text{ below} = P(Z)$$
$$\% \text{ OK} = 100 - (\% \text{ above} + \% \text{ below}) \tag{4.8}$$

It is apparent that if the manufacturing process is not centered with the specification nominal (second case in the table), the total defect rate increases, even if the manufacturing process standard deviation remains the same. Similar increases in the defect rate occur if the manufacturing process standard deviation increases or there is a comparable decrease in the tolerance limits of the design. Table 4.2 also illustrates the use of C_p or C_{pk} depending on whether the manufacturing process mean is centered around the design specification nominal.

4.2.2 Determining the Defects for a Multistep Manufacturing Process

Manufacturing is generally a multistep process. Each step generates its own variability and therefore contributes to the overall reject rate. In a large multistep operation, the individual process quality has to be very high; if it is not, the overall yield will be low. If there is no in-process inspection or test to cull out the intermediate rejects, the final reject rate could be very high, as is the case in PCB and IC fabrications, which have 30 to 50 production steps.

It is important to measure quality in terms of the total number of defects found anywhere in the manufacturing process, and prior to any test or inspection. This will reduce confusion when setting quality targets or comparing similar operations in different plants. In addition, it is a true measure of quality, which is not masked by the test or inspection costs.

Units of these quality measures are expressed in terms of defects per unit (DPU), first-time yield (FTY) and parts per million (PPM). They are more accurate than the traditional reject rates, especially when repairs are not considered as part of the acceptable product category. The following are the equations used to describe these units and their relationships:

$$\text{DPU} = \frac{\text{number of defects found anywhere}}{\text{number of units processed}} \tag{4.9}$$

$$\text{PPM} = \frac{\text{DPU}}{1{,}000{,}000} \tag{4.10}$$

$$\text{FTY} = e^{-\text{DPU}} \tag{4.11}$$

MANUFACTURING VARIABILITY AND PRODUCT SPECIFICATIONS 77

PPM is the normalization of the DPU by a factor of 1,000,000 in order to facilitate equating a lower number with lower rejects and driving it down to zero. The first-time yield (FTY) is presented as a Poisson distribution, of mean equal to DPU, since the defects are assumed to occur in a random fashion during the process. The advantages of treating the yield as Poisson distribution is that it can very quickly estimate the FTY in a multistep process by adding the DPUs. For example, a three-step process, A, B, and C, the FTY calculations would be as follows:

Process step	A	B	C
DPU at process step	a	b	c
Process yield (FTY)	e^{-a}	e^{-b}	e^{-c}
Total process yield $Y\{TOT\} =$	$Y\{A\} \times Y\{B\} \times Y\{C\}$		
FTY $\{TOT\}$ $=$	$e^{-(a+b+c)}$		

4.2.3 Example: Determining First-Time Yield at the Electronic Product Turn-on Level

The new electronic products being developed today are more complex than previous products. The number of components on each PCB is increasing, as well as the total number of PCBs in the product. In this example, the effects of these complexities on the final product turn-on will be demonstrated. The historical quality level that sustained the production process for older products will be inadequate for the new complex products. The in-process manufacturing quality of components and PCBs will have to be improved significantly to counteract the increased number of assemblies and components.

The defect rate for new PCBs is calculated on the basis of process observations for existing PCBs. Assuming a through-hole technology, defects are usually obtained from three sources: incoming materials and components; assembly defects of missing, wrong or reversed components; and soldering defects. If it is assumed that each component has 2.5 connections per PCB, the quality level for multiple-components PCBs can be calculated as follows, assuming reasonable current PCB assembly process quality.

Example at PCB Assembly Level

Assume current defect rates at the PCB process:

Solder defect rate $= 100$ ppm
Component assembly defect $= 500$ ppm
Component defectives $= 300$ ppm
Connections per component 2.5

Total process yield at board test level for 100, 500 and 1000 component PCBs:

$$FTY\{TOT\} = e^{-\{(SOLDER \times \#PARTS \times 2.5) + ASSY + DEF\}}$$

# PARTS	SOLDER	ASSY	DEF	TOT DPU	YIELD
100	0.025	0.05	0.03	0.105	90%
500	0.125	0.25	0.15	0.525	60%
1000	0.25	0.5	0.3	1.05	35%

It can be shown that as the number of components increase in the PCBs, first-time yields decrease significantly if the quality level of the assembly process remains the same. In order to achieve higher first-time yields for PCBs, the quality level of the assembly process has to be improved from PPM of defects in the hundreds to PPM defects in the tens.

Example at PCB Auto Test Level

Usually, these PCBs are tested on automatic in-process testers, which removes some of these defects. Final assembly of the electronic product is accomplished from these tested PCBs and other components such as power supplies and display devices and turned on for final test. Similar calculations hold true for the first-time yield for the product at final test.

If the product contains 10 PCBs, and a goal of 95% turn-on yield was set for each PCB at auto test, the product final test turn on yield will be:

$$DPU = 0.05 \tag{4.12}$$

$$FTY\{TOT\} = e^{-10 \times (0.05)} = 60\% \tag{4.13}$$

A turn-on yield of 60% is disappointing, especially when 95% in process PCB yield is difficult to achieve using today's auto test technology. To achieve a 95% final product turn on, a DPU of 0.005 is required, for an individual PCB test yield of 99.5%:

$$\begin{aligned} FTY\{TOT\} &= e^{-10 \times (DPU)} = 95\% \\ 10 \times DPU &= -\log(0.95) = 0.05 \\ DPU &= 0.005 \\ PCB\ &individual\ test\ yield = 99.5\% \end{aligned} \tag{4.14}$$

The manufacturing process has to increase its quality level to meet the challenge of increased complexity. Several methodologies and tools can be used for each part of the assembly process. These steps do not necessarily imply the use of more sophisticated inspection methods and equipment, but rather such simple problem solving techniques as the following. *Incoming component* quality can be improved with better supplier certification and supplier process control methods. *Assembly process* quality can be enhanced with better employee training, the use of more automation such as auto-insertion and auto-placement of components, and improvement of the design guidelines for PCBs such as component polarity indicators, components placement and orientation in one axis, and graphic placement aids. *Soldering quality* can be improved by continuously upgrading soldering materials and processes with the latest technology available, and performing design of experiment techniques to optimally meet the soldering process parameters.

4.3 MANUFACTURING VARIABILITY MEASUREMENT AND CONTROL

Control charts have traditionally been used as the method of determining the performance of manufacturing processes over time by the statistical characterization of a measured parameter that is dependent on the process. They have been used effectively to determine whether the manufacturing process is in statistical control. Control exists when the occurrence of events (failures) follows the statistical laws of the distribution from which the sample was taken.

Control charts are run charts with a center line drawn at the manufacturing process average and lines drawn at the tail of the distribution at the 3σ points. If the manufacturing process is under statistical control, 99.74% of all observations are within limits of the process. Control charts by themselves do not improve quality. They merely indicate that the quality was in statistical "synchronization" with the quality level at the time when the charts were created.

There are two major types of control charts: variable charts that plot continuous data from the observed parameters; and attribute charts, which are discrete and plot accept/reject data. Variable charts are known as X-bar and R charts. Figure 4.3 is a typical chart showing the solder paste height deposition process for an SMT (surface mount technology) process. Attribute charts tend to show proportion or percentage defective. There are four types of attribute charts, P chart, NP chart, C chart and U chart. Their usage is shown in Table 4.3.

The selection of the parameters to be control-charted is an important part of the design for manufacture process. Too many parameters plotted tend to adversely confuse the beneficial effect of the control charts, since they will all move in the same direction when the process is out of control. It is very important that the parameters selected for control charting be independent from each other, and directly related to the overall performance of the product.

When a chart shows an out-of-control condition, the process should be investigated and the reason for the problem identified on the chart. Figure 4.4 shows a typical scenario of plotting a parameter (in this case the surface cleanliness measurements on PCBs) which was due to a defective laminate lot. Note that the new lot has significantly increased the resistance value.

When introducing control charts to a manufacturing operation it is beneficial to use elements that are universally recognized, such as temperature

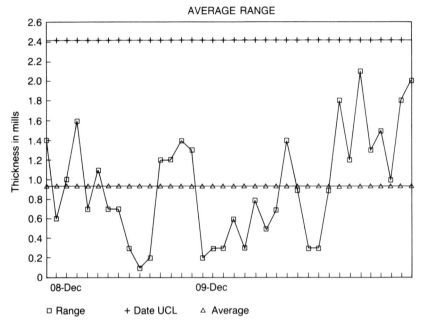

Figure 4.3. Typical X-bar and R chart for SMT solder paste height.

and relative humidity, or to take readings from a process display monitor, such as the temperature indicator in a soldering system. Since the data collection is simplified, these control charts can be used to introduce and train the operators in data collection and plotting of parameters.

When introducing control charting to an organization, the production operators have to be directly active in the charting process to increase their awareness and get them involved in the quality output of their jobs. Several

TABLE 4.3. Control Charts for Attributes

Type of counting		Data plotted	Subgroup or sample size
Partitioned counts	Featured counts		
Chart type			
NP	C	Integers	Must be constant
P	U	Fractions or rates	Can vary

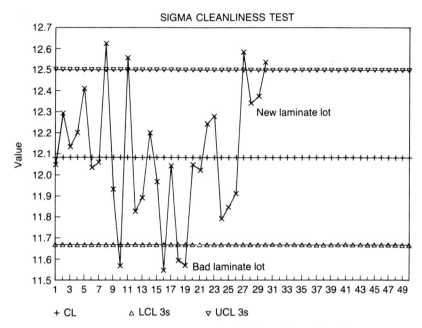

Figure 4.4. Out-of-control condition for surface cleanliness.

shortcomings have been observed when first using control charts, some of those to avoid are:

- Improper training of the production operators: Collecting a daily sample and calculating the average and range of the sample data set might seem a simple enough task. Unfortunately, because of the poor set of operator skills in many manufacturing plants, extensive training has to be given to make sure the manufacturing operator can perform the data collection and calculation tasks for control charting.
- Using one of the software tools for plotting data removes the focus from the data collection and manipulation of control charting and focuses it on operating the software tools. Automatic means of plotting control charting should be done later in the control charting plan for production.
- Selecting variables that are outside of the production group's direct sphere of influence, or are difficult or impossible to control, could result in a negative perception of the quality effort. Examples are plotting temperature and relative humidity in a poorly designed ventilation plant, where the control charts could continuously show an out of control situation that is expensive and difficult to correct.

4.3.1 Generation of Control Charts

The X-bar chart shows where the manufacturing process is centered. If the process is stable, the chart is centered. If there is a trend in the plotted data, then the process value is moving up or down. Causes for the X-bar chart to be moving include faulty machine or process settings, biased operators, and bad materials.

The R chart shows the uniformity or consistency of the manufacturing process. If the R chart is narrow, then the product is uniform. If the R chart is wide or out of control, then there is a nonuniform effect on the process, such as poor repair or maintenance record, new operators or nonuniform materials.

The charts are generated by taking a historical record of the manufacturing process over a period of time. Shewhart, the father of control charts, recommends that "statistical control cannot be reached until under the same conditions, not less than 25 samples of four each have been taken to satisfy the required criterion" (Shewhart, 1931). These observations form the historical record of the process. All observations from now on are compared to this baseline.

From these observations, for each sample set, the sample set average $\bar{X}$ and the sample range R (highest value − lowest value) are recorded for that time period. At the end of the observation period (minimum 25 samples), the average of $\bar{X}$ values, designated as $\bar{\bar{X}}$, and the average of R values, designated as $\bar{R}$, are recorded. The control limits for the control charts are then calculated using the following formulas and control chart factors from Table 4.4.

X control limits (3-sigma limits):

$$\text{Upper control limit (UCL)} = \bar{\bar{X}} + 3\sigma$$
$$\text{Lower control limit (LCL)} = \bar{\bar{X}} - 3\sigma$$

since

$$3\sigma = A_2 \bar{R} \qquad \text{UCL}_x = \bar{\bar{X}} + A_2 \bar{R}, \qquad \text{LCL}_x = \bar{\bar{X}} - A_2 \bar{R}$$

R control limits:

$$\text{Upper control limit (UCL)}_R = D_4 \bar{R}$$
$$\text{Lower control limit (LCL)}_R = D_3 \bar{R}$$

$\bar{X}$ = average of n observation in a subgroup
$\bar{\bar{X}}$ = average of all X's
R = range of n observation in a subgroup (highest − lowest value)
$\bar{R}$ = average of all R's
A_2 = factor for X chart
D_3 = lower control limit factor for R chart
D_4 = upper control limit factor for R chart

TABLE 4.4. Control Chart Factors

Number of observations in subgroup, n	Factor for $\bar{X}$ chart, A_2	Factors for R chart		Factor d_2, $d_2 = \bar{R}/\sigma$
		Lower control limit D_3	Upper control limit D_4	
2	1.88	0	3.27	1.128
3	1.02	0	2.57	1.693
4	0.73	0	2.28	2.059
5	0.58	0	2.11	2.326
6	0.48	0	2.00	2.534
7	0.42	0.08	1.92	2.704
8	0.37	0.14	1.86	2.847
9	0.34	0.18	1.82	2.970
10	0.31	0.22	1.78	3.078
11	0.29	0.26	1.74	3.173
12	0.27	0.28	1.72	3.258
13	0.25	0.31	1.69	3.336
14	0.24	0.33	1.67	3.407
15	0.22	0.35	1.65	3.472
16	0.21	0.36	1.64	3.532
17	0.20	0.38	1.62	3.588
18	0.19	0.39	1.61	3.640
19	0.19	0.40	1.60	3.689
20	0.18	0.41	1.59	3.735

Control chart limits indicate a different set of conditions from the specification limits. If they are equal to each other, then the process capability C_p equals 1, and the reject rate is found to be 0.26% or 2,600 PPM, from the normal probability distribution curves for a Z-value of 3. It is desirable to have the specification limits as large as possible when compared to the process control limit.

The control limits represent the 3σ point, based on a sample of n observations. To determine the standard deviation of the product population, the central limit theorem can be used:

$$s = \sigma/\sqrt{n} \qquad (4.15)$$

where
s = sample deviation
σ = population deviation
n = sample size

84 PRODUCT SPECIFICATIONS AND MANUFACTURING PROCESS TOLERANCES

Multiplying the distance from the center line of the X-bar chart to one of the control limits by $\sqrt{n}$ will determine the total product population deviation. A simpler approximation is the use of the formula $\sigma = R/d_2$ in Table 4.4 to generate the total product standard deviation directly from the control chart data.

Example: Variable Control Chart Calculations

In this example, a critical dimension in a part is measured as it is being inspected in a machining operation. To set up the control chart, four measurements were taken every day for 25 successive days, to approximate the daily production variability. These measurements are then used to calculate the limits of the control charts. The measurements are shown in Table 4.5.

During the first day, four samples were taken, measuring 19, 21, 22, and 18 thousandths of an inch. These were recorded in the top of the four columns of sample #1. The average, or $\bar{X}$ value, was calculated and entered in column 5:

$$\bar{X} = (19 + 21 + 22 + 18)/4 = 80/20 = 20$$

The range, of R value, is calculated by taking the highest reading (22 in this case), minus the lowest reading (18 in this case):

$$R = 22 - 18 = 4$$

The averages of $\bar{X}$ and R are calculated by dividing the column totals of $\bar{X}$ and R by the number of subgroups:

$$\bar{\bar{X}} = (\text{sum of } \bar{X})/(\text{number of subgroups})$$
$$\bar{\bar{X}} = 487.50/25 = 19.5$$
$$\bar{R} = (\text{sum or } R)/(\text{number of subgroups})$$
$$\bar{R} = 116/25 = 4.64$$

Using the control chart factor table (Table 4.4), the control limits can be calculated, using the number of observations in each subgroup (4) as the row number:

$$\text{UCL}_x = \bar{\bar{X}} + A_2\bar{R} = 19.5 + (0.73 \times 4.64) = 22.89$$
$$\text{LCL}_x = \bar{\bar{X}} - A_2\bar{R} = 19.5 + (0.73 \times 4.64) = 16.11$$

R control limits:

$$\text{Upper control limit } (\text{UCL})_R = D_4\bar{R} = 2.28 \times 4.64 = 10.58$$
$$\text{Lower control limit } (\text{LCL})_R = D_3\bar{R} = 0$$

Since the measurements were recorded in thousandths of an inch, the center line of the $\bar{X}$ control chart is 0.0195 and the control limits for X are 0.02289 and 0.01611. For the R chart, the center line is set at 0.00464 and the limits are 0.01058 and 0.

These numbers form the limits of the control chart. Each day, four readings of the part decision are to be taken by the responsible operators, with the average of the four readings plotted on the $\bar{X}$ chart, and the range or difference between the highest and lowest reading to be plotted on the R chart.

TABLE 4.5. Control Chart Limit Calculations Example. Basic unit of measure = 0.001 inch

Sample No.	Parts				Average $\bar{X}$	Range R
	1	2	3	4		
1	19	21	22	18	20.00	4
2	21	19	18	20	19.50	3
3	22	16	21	20	19.75	6
4	17	18	19	20	18.50	3
5	24	18	22	21	21.25	6
6	17	18	19	22	19.00	5
7	18	20	21	18	19.25	3
8	19	17	18	21	18.75	4
9	15	19	23	22	19.75	8
10	22	23	20	21	21.50	3
11	16	16	21	20	18.25	5
12	19	15	22	17	18.25	7
13	22	21	16	19	19.50	6
14	21	21	18	17	19.00	4
15	17	19	16	20	18.00	4
16	17	18	19	22	19.00	5
17	18	20	21	18	19.25	3
18	19	17	18	21	18.75	4
19	15	19	23	22	19.75	8
20	22	23	20	21	21.50	3
21	19	21	22	18	20.00	4
22	21	19	18	20	19.50	3
23	22	16	21	20	19.75	6
24	17	18	19	20	18.50	3
25	24	18	22	21	21.25	6
Totals					487.50	116

These control limits can now be used for future production days to monitor whether the process remains in control. Using the Table 4.4 to estimate the standard deviation σ from R,

$$\sigma = \bar{R}/d_2 = 4.64/2.059 = 2.25 \text{ mils}$$
$$\pm 3\sigma = \pm 6.75 \quad \text{or} \quad 0.00675 \text{ inch} \quad (4.16)$$

The 3σ limit indicates that 99.74% of the measurements should fall within ± 0.00675 inch. If the specified tolerance was 0.020 ± 0.010 then the process capability index would be

$$C_p = (U - L)/6\sigma = (0.030 - 0.010)/(6 \times 0.00225) = 0.020/0.0135 = 1.5 \quad (4.17)$$

4.3.2 Interpretation of Control Charts

Control charts are not directly related to defects. Control limits represent the state of the manufacturing process at the time the chart was created, and will indicate when the process has gone beyond those original conditions. The defects are generated only when the product measurements exceed the design tolerance specifications. For a robust manufacturing process, the control limits should be observed to be much smaller than the design tolerance specifications.

When a process is in statistical control, the normal pattern is to have the point plotted for each subgroup to be near the center, with a few points near the control limits, but rarely outside them (see Figure 4.5).

An out-of-control pattern is an indication that there is a system change that merits investigation. Out-of-control problems can be detected by observing the plotted points and where they are located by dividing each half of the control chart into three equal zones, and considering the top and bottom separately (see Table 4.6).

Other patterns of in-control charts that should trigger investigations, even if the points plotted are not outside the control limits, are trends such as increasing average, change in variance, decreasing runs, and cycling of the parameter being measured. These are illustrated in Figures 4.6 through 4.9.

4.3.3 Tools of Continuous Process Improvements

Control charts are one of the elements used in improving the quality of the manufacturing process. They serve as good baseline indicators, and as a reporting mechanism for the quality improvement efforts. Improving the

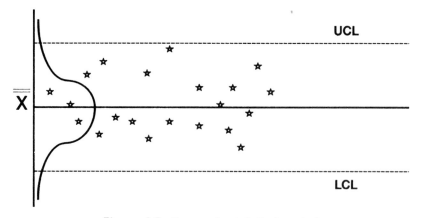

Figure 4.5. Process in statistical control.

MANUFACTURING VARIABILITY MEASUREMENT AND CONTROL

TABLE 4.6. Interpretation of Control Charts

```
----------- UCL
       A                        (1) One point outside control limit
  - - - - - -
       B
  - - - - - - -                 (2) 2 of 3 successive points zone A or
                                    beyond
       C
----------- CENTER
       C                        (3) 4 of 5 successive points zone B or
                                    beyond
  - - - - - -
       B
  - - - - - - -                 (4) 8 successive points zone C or beyond
       A
----------- LCL
```

quality of the manufacturing process through constant efforts at identifying and eliminating the causes of defects is the key element of success, with control charts acting as the scorekeeper for continuous process improvements. The following set of tools which were developed for continuous process improvements have been used successfully through different companies and organizations. They include techniques for collecting defect data, manipulating

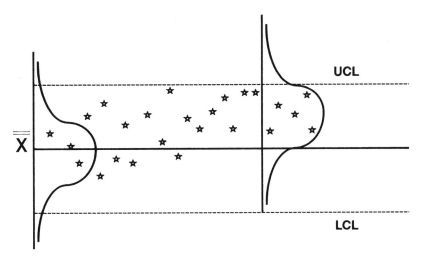

Figure 4.6. Process out of control—increasing average.

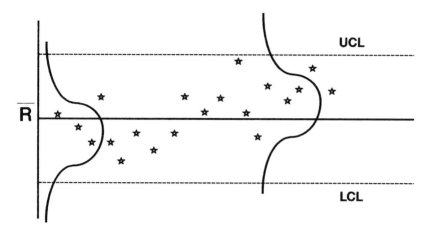

Figure 4.7. Process out of control—change in variance.

and plotting them, prioritizing and identifying defect causes, and removing the most probable ones.

These tools of quality are used in other concurrent engineering techniques such as design of experiments and quality function deployment.

Brainstorming

Brainstorming is a technique used to get a group to generate as many ideas as possible on a topic or a problem. Brainstorming is useful to open

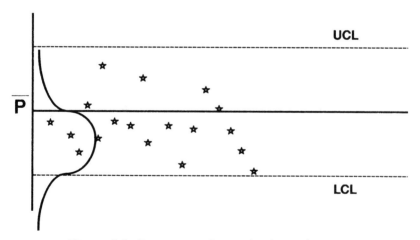

Figure 4.8. Process out of control—decreasing run.

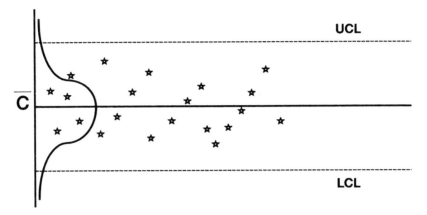

Figure 4.9. Process out of control—cycling.

discussions by involving all group members to generate as many ideas as possible without bias to any single idea.

Brainstorming is a good tool to use for group discussion in trying to solve a problem or initiate an action. It has been used extensively in developing and focusing teams of engineers to solve problems or generate ideas for completing and initiating tasks.

The group members should be knowledgeable on the topic to be discussed. Every member should participate in brainstorming. The ideas should be recorded promptly without any arguments and no one person should dominate the team.

Phases of brainstorming

Idea generation

- Toss out as many ideas as possible; list these ideas on flip chart or sticky-sided papers.
- All ideas are permitted; be as freewheeling as possible. One good idea can trigger another.
- Do not interrupt or analyze; there should be no jumping to conclusions. Ask questions only to clarify recording.
- Adapt or build on ideas already listed.

Clarification

- Repeat all items on the list and have everyone agree and understand each idea.
- Remove duplications, and add any new ideas.
- Record the list as necessary.

90 PRODUCT SPECIFICATIONS AND MANUFACTURING PROCESS TOLERANCES

Evaluation
- Narrow down the list by allowing discussions.
- Agree on a final list of ideas acceptable to the group.

It is advisable to use simulated training sessions for brainstorming. A group could attempt to tackle a problem, such as the design of a paper airplane or improving a golf swing or a tennis game, before embarking on brainstorming the problem at hand.

Cause and Effect Diagrams

This tool shows the relationship between the effect (rejects) and its possible causes. It is used to logically group and identify all possible problems. It is also referred to as the "fishbone" or "Ishikawa" diagram.

Constructing a cause and effect diagram
- Use brainstorming to identify all possible causes for the effect. Ask outside experts to add to the list produced by brainstorming.
- Review the list and look for any interrelationships between the possible causes. Define three to six major categories that can be grouped together, and categorize them. Common categories are sometimes referred to as the four Ms: materials, machines, methods, and manpower.
- Within each category, further subdivision might be required based on relationship or cause. They can be ultimately divided into subgroups.
- Draw the diagram, using arrows and means of each group, subgroup, and individual cause (see Figure 4.10).
- Evaluate and select the most probable cause(s) on the basis of the problem-solving group decision.

Once the most probable cause has been identified, problem-solving techniques such as design of experiments can be used to verify the problem cause and institute corrective action.

Checksheets

A checksheet is a form used to identify, gather, organize, and evaluate data. A well-designed checksheet can eliminate confusion, enhance accuracy, and reduce time needed to take data.

TYPES OF CHECKSHEETS

RECORDING CHECKSHEET (Figure 4.11) This is used for recording data on types of defects. The types should be listed in categories, and a mark is made each time a defect is found in the sample. It is important not to collect too many types of defects.

MANUFACTURING VARIABILITY MEASUREMENT AND CONTROL 91

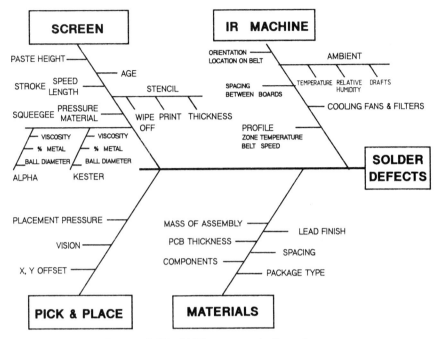

Figure 4.10. SMT cause and effect diagram.

PROCESS _____   DATE _____
SAMPLE RATE _____   SAMPLE SIZE _____
PCB TYPE _____   DEPT _____

DEFECT TYPE	CHECK	SUBTOTAL
DEWETTING		7
EXCESS SOLDER		11
SOLDER BRIDGE		18
OTHER		9
	TOTAL	45

Figure 4.11. Defect data recording checksheet.

92 PRODUCT SPECIFICATIONS AND MANUFACTURING PROCESS TOLERANCES

It is difficult to properly train production operators to distinguish between very similarly worded types of defects, even if photographs and other methods of graphically presenting them are used. Count the total number of checks for each defect.

LOCATION CHECKSHEET (Figure 4.12) This is used to collect the location of the defects, and how often they occur. This technique is useful for identifying concentration of defects on a printed circuit board.

Other information should be included when available, such as date, part number, lot number, supplier name, supplier date code, area location, and so on. Using automatic means of collecting and categorizing data such as bar-code readers and scanners can speed up the recording of data. The defects data categories can be arranged in the bar-code format so that an operator with a bar-code wand can enter all the data without writing down any information by hand.

Flowcharts

A flowchart is a picture of the process. It represents a step-by-step sequence and can help in reaching a common understanding on how the manufacturing

```
PROCESS _____        DATE _____
SAMPLE RATE _____        SAMPLE SIZE_____
PCB TYPE _____        DEPT _____
```

DEFECT TYPE

A - UNFILLED HOLES   D - REVERSE COMPONENT   G - PCB DELAMINATION
B - EXCESS SOLDER   E - MISSING COMPONENT   H - OPEN FOIL
C - SOLDER BRIDGE   F - WRONG COMPONENT   I - OTHER

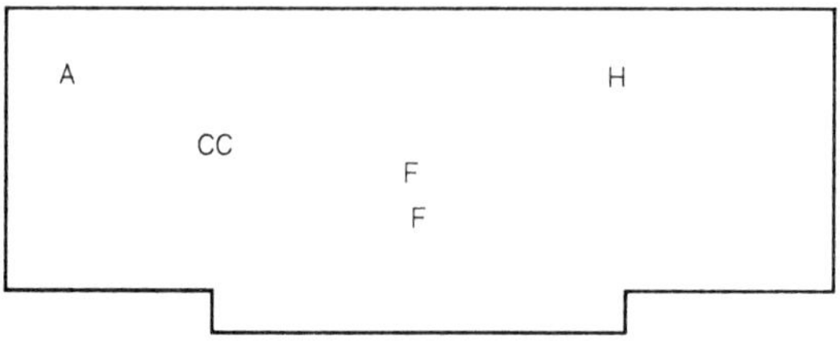

```
   A                        H
      CC
              F
              F
```

Figure 4.12. Defect data location checksheet.

process is run, and can act as a base for enhancing or changing the process. It can also be used as a documentation and training tool, for pointing out areas for data collection and control, and can be used as a brainstorming basis for enhancing and troubleshooting the manufacturing process. Figure 4.13 is a flowchart representation of control charts.

Flowcharting process

- Identify first and last step of the process.
- Fill in each process step. Include any time the product is handled, transferred, joined, or changed in form.
- Show feedback loops, such as rework paths: They indicate inefficiency and possible low quality.
- Choose symbols that are well understood or have been previously used: Oblong for start/end of process; diamonds for steps; squares for decision points.
- Use structured approach to simplify charts. Break down each major step into a box used in upper-level chart. Make sure all lines connect.
- Keep charts up to date as process evolves.

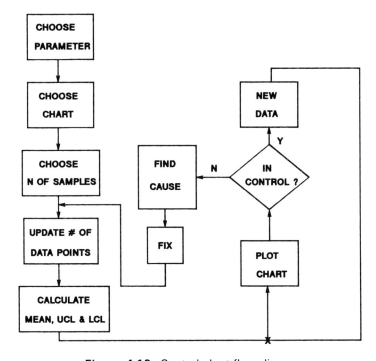

Figure 4.13. Control chart flow diagram.

94 PRODUCT SPECIFICATIONS AND MANUFACTURING PROCESS TOLERANCES

Pareto Charts

Pareto charts have data plotted in a bar graph form and display the number of times each defect has occurred, in ascending order. They plot the relative contribution of each defect cause, and tell at a glance the largest causes.

Pareto charting process
- Decide how many categories to plot. This will be equal to the total number of bars.
- Draw horizontal (X) axis. Label each category. Draw vertical axis with percentage and total number of occurrences for each category shown for each bar.
- Use bars of the same width arranged from tallest to shortest.
- Add information: title, preparer, date, and so on.

The Pareto principle is similar to the "80–20" rule: 20% of the problems cause 80% of the defects. It could be used to focus on the right causes of defects, as well as to prioritize them. Ideally, a Pareto chart should have all small bars!

Figure 4.14 is a Pareto chart presentation of a PCB fabrication shop, showing the relative distribution of defect sources in terms of their occurrences.

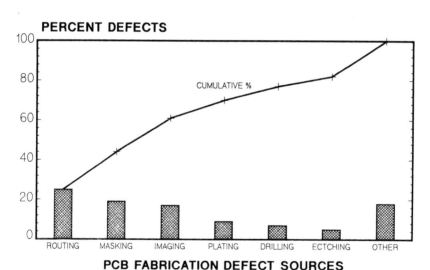

Figure 4.14. Pareto chart of PCB fabrication process.

Scatter Diagrams

Scatter diagrams are simple graphical methods used to study relationships between two variables. They can quickly determine whether a relationship (positive or negative) exists and the strength (correlation) of that relationship.

Scatter diagram procedures

- Decide how many points to plot. A minimum of 25 points are needed to make conslusions significant.
- Arrange the pairs of measurements in ascending value of x. Divide data into subgroups of x.
- Draw and label horizontal and vertical axes. Choose the proper scale to fit all points.
- If the diagram shows an upward trend, there is a positive correlation. A downward trend is negative, and a level trend implies no correlation between variables.
- It might be necessary to plot logarithmic scales or many y points to a single x point to show data.

Use regression analysis to accurately determine degree of correlation and the "line of best fit." Use any available software (calculators included) to determine significance ($t > 3$), using the formula:

$$t(\text{number of } \sigma\text{'s}) = r \times \sqrt{n} \qquad (4.18)$$

where r is the correlation factor and n is the number of data pairs.

Histograms

Histograms are a picture of the frequency distribution of the process. They are bar graphs with the height of each column representing the number of occurrences in each step for the process or measurement.

By drawing the process specification on the axis, histograms clearly show the position of the process relative to desired performance. It becomes clear whether the process is performing as desired, or whether the process average needs to be shifted, or the distribution needs to be narrowed.

One of the problems in plotting histograms is determining the best fit for a probability distribution. The following method will help determine the best cell boundary to match a possible distribution.

Fitting histogram curve to data

1. Calculate data average $\bar{X}$ and standard deviation σ.
2. Translate data cell boundary in terms of σ, $(B - \bar{X})/\sigma$, where B is the data cell boundary.

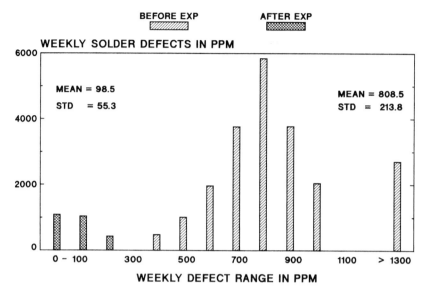

Figure 4.15. Solder process quality 6-month histogram.

3. Look up each cell boundaries pair for areas under the curve, and determine the possible theoretical distribution that will match the data. The difference in areas equals the theoretical percentage for each cell.
4. Draw a smooth line through each cell to show the distribution. Use values from step (3) above.
5. Plot observed data and normalize frequencies into observation percentage = observed/total.

Figure 4.15 is a histogram presentation of data for the improvement of a PCB soldering process.

Time Series Graphs

Time series graphs are sometimes called run charts. They are line charts used to monitor process quality measures over time. Run charts identify how process parameters change with time and indicate trend, shifts, and process cycling. They should be used to set quality process measures and goals.

Information from run charts

- Decide on quality units: A universal one such as defect per unit (DPU) or parts per million (PPM) can be used. PPM goals are universal. Compare with similar process in other companies or locations. Examples

MANUFACTURING VARIABILITY MEASUREMENT AND CONTROL   97

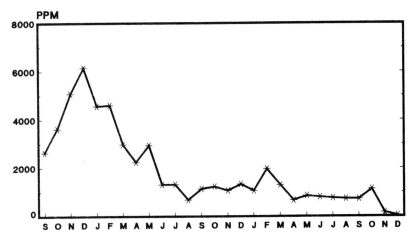

Figure 4.16. Solder process monthly defects run chart.

are solder shorts per million solder connections, and missing PCB holes per million holes drilled.
- Show goal line if appropriate. These goals should be set aggressively. However, do not set them if they are impossible to meet. Realistic goals should be reached first, then they can be set higher when current ones are met.

Figures 4.16 and 4.17 are run charts for a soldering process, showing the performance of the process quality and uptime over a period of 2 years.

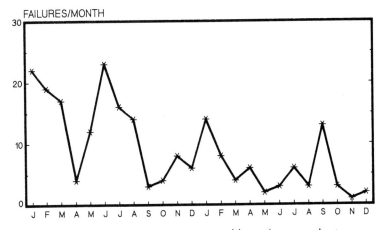

Figure 4.17. Solder process monthly uptime run chart.

4.4 SETTING THE PROCESS CAPABILITY INDEX

Many companies are beginning to think about the process capability index C_p as a good method for both design and manufacturing engineers to achieve quality goals jointly, by having both sides work together. Design engineers should open up the specification to the maximum possible, while permitting the product to operate within customer expectations. Manufacturing engineers should reduce the process variations by maintenance and calibration of processes and materials, and by performing design of experiments.

Another advantage of using the C_p as a quality measure and target is the involvement of the suppliers in the design and development cycle. To achieve the required C_p, the design engineers must know the quality level and specification being delivered by the suppliers and their materials and components. In some cases, the suppliers do not specify certain parameters, such as rise time on integrated circuits, but provide a range. The design engineers must review several samples from different lots from the approved supplier and measure the process variability based on those specifications. A minimum number of 25 samples is recommended.

Many companies target a specific C_p level to set expected design specifications and process variability targets for each part or assembly. This C_p level predetermines the specification limits as a measure of process capability. Usually this number has been used to set a particular defect rate such as 100PPM, which is a C_p of 1.3 and a specification limit of $\pm 4\sigma$. Motorola Incorporated has set as corporate goal a C_p of 2, a defect rate of 2 parts per billion or a specification limit of $\pm 6\sigma$. This process capability is sometimes referred to as six-sigma. A summary of the different levels of C_p and their sample and defects rates is given in Table 4.7.

Another advantage of a high C_p value is that a process mean shift does not affect the reject rate. A C_p of 2 signifies a very robust process where a

TABLE 4.7. The Effects of C_p on Sample and Defect Rates

C_p	Percent yield	Defect rate %	Defect rate PPM	Number sampled to fine one defect
0.86	99.00	1	10,000	100
1.0	99.74	0.26	2,600	260
1.3	99.99	0.01	100	10,000
1.5	99.999	0.0001	10	100,000
2.0	99.9999998	0.0000002	2 PPB	500,000,000

SETTING THE PROCESS CAPABILITY INDEX

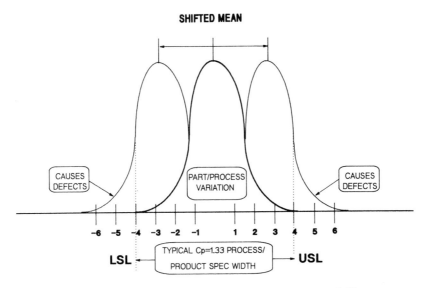

Figure 4.18. Typical defect rate for shifted mean with $C_p = 1.33$.

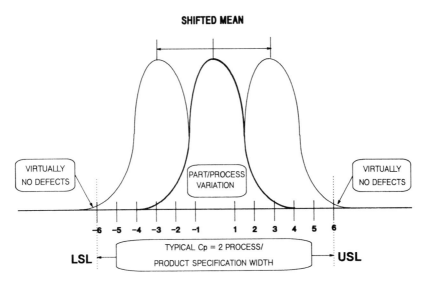

Figure 4.19. Typical defect rate for shifted mean with $C_p = 2$.

shift of the mean by as much as 1.5σ imparts a defect level of 3.4 PPM to the end product. A comparable shift of the mean for a C_p of 1.33 increases the defect rate to 6800 PPM (see Figures 4.18 and 4.19).

4.5 CONCLUSION

It can be seen from this chapter that the power of the process capability index is the cooperative joining of responsibility for quality improvements between the manufacturing and design engineering departments. The design engineering department's responsibility is to set the specification limits for new products as broadly as possible and still permit the proper functioning of the product. The manufacturing engineering department has to narrow the manufacturing process distribution, as measured by the standard deviation of the product characteristic. This can be achieved by more frequent maintenance schedules, improvement of incoming inspection methods, working with suppliers, and performing design of experiments to reduce the variability of the process.

The process capability index specifically implies the use of quality methods in manufacturing such as statistical process control charts and the quality tools used to reduce variability and improve quality.

The process quality improvement tools of brainstorming, cause and effect diagrams, scatter diagrams, Pareto diagrams, checksheets, flowcharts, histograms, and control charts can be used effectively to enhance the process and increase the quality level by reducing variability.

SUGGESTED READING

AT&T. *Statistical Quality Control Handbook*. Easton, Pa.: Mack Printing Company, 1984.

Afifi, A. and Azen, S. *Statistical Analysis; A Computer Oriented Approach*, 2d edn. New York: Academic Press, 1979.

American National Standards Institute (ANSI). "Control charts methods of analyzing data." *ASQC Standard B2/ANSI 21.2*.

American National Standards Institute (ANSI). "Control charts methods of controlling quality during production." *ASQC Standard B3/ANSI 21.3*.

American National Standards Institute (ANSI). "Guide for quality control charts." *ASQC Standard B1/ANSI 21.1*.

American National Standards Institute (ANSI). *ANSI/IPC-PC-90 Standard*. Developed by the IPC, Lincolnwood, Ill.

American Society for Quality Control (ASQC). "Definitions, symbols, formulas and tables for control charts." *ANSI/ASQC A1*.

SUGGESTED READING

Anderson, V. and McLean, R. *Design of Experiments.* New York: Marcel Dekker, 1978.
Bowker, A. and Lieberman, G. *Engineering Statistics.* Engelwood Cliffs, N.J.: Prentice-Hall, 1972.
Box, G. and Hunter, W. *Statistics for Experimenters.* New York: Wiley, 1978.
Burr, I. *Engineering Statistics and Quality Control.* New York: McGraw-Hill, 1953.
Chan, Lai et al. "A new measure for process capability: C_{pm}." *Journal of Quality Technology.* Vol. 20, No. 3, 1988, pp. 162–175.
Clausing, D. and Simpson, H. "Quality by design." *Quality Progress* (Journal of the ASQC), January 1990, pp. 41–44.
Cowden, Dudley J. *Statistical Methods in Quality Control.* Engelwood Cliffs, N.J.: Prentice-Hall, 1957.
Crosby, P. *Quality is Free.* New York: McGraw-Hill, 1979.
Deming, Edwards. *Quality, Productivity and Competitive Position.* MIT Center for Advanced Engineering Studies, 1982.
Devore, Jay. *Probability and Statistics for Engineering and the Sciences.* Belmont, Calif.: Brooks/Cole, a Division of Wadsworth Co., 1987.
Dixon, W. and Massey, F. *Introduction to Statistical Analysis.* New York: McGraw-Hill, 1969.
Duncan, Acheson J. *Quality Control and Industrial Statistics,* 4th edn. Homewood, Ill.: Richard D. Irwin, 1986.
Feigenbaum, A.V. *Total Quality Control,* 3d edn. New York: McGraw-Hill, 1983.
Gill, Mark S. "Stalking six sigma." *Business Month Journal,* January 1990.
Grant, E. and Leavenworth, R. *Statistical Quality Control,* 5th edn. New York: McGraw-Hill, 1980.
Hahn, G. and Shapiro, H. *Statistical Models in Engineering.* New York: McGraw-Hill, 1967.
Harry, M.J. *The Nature of Six Sigma Quality.* Motorola Incorporated, 1988.
Heindreich, Paul. "Designing for manufacturability." *Quality Progress* (Journal of the ASQC), May 1988.
Ishikawa, K. *Guide to Quality Control* (revised edn.). Asian Productivity Institute, 1976.
Jurna, J. and Gryna, Frank. *Quality Control Handbook,* 4th edn. New York: McGraw-Hill, 1979.
Jurna, J. and Gryna, Frank. *Quality Planning and Analysis.* New York: McGraw-Hill, 1970.
Kane, V. "Process capability indices." *Quality Technology Journal.* Vol. 18, pp. 41–52, 1986.
Kendrick, J. "Hewlett Packard quest for quality." *Quality Journal,* November, pp. 16–20, 1988.
King, James. *Probability Charts for Decision Making.* New York: The Industrial Press, 1971.
Miller, Irwin and Freund, John E. *Probability and Statistics for Engineers.* Engelwood Cliffs, N.J.: Prentice-Hall, 1965.
Moran, J., Talbot, R. and Benson, R. *A Guide to Graphical Problem Solving Processes.* Milwaukee, Wis.: ASQC Press.
Nelville, A. and Kennedy, J.B. *Basic Statistical Methods for Engineers and Scientists.* New York: International Textbook Company, 1964.
Ott, E.R. *Process Quality Control.* New York: McGraw-Hill, 1975.
Ott, E.R. *An Introduction to Statistical Methods and Data Analysis.* North Scituate: Duxbury Press, 1977.
Ryan, Thomas. *Statistical Methods for Quality Improvements.* New York: Wiley, 1989.

Shina, S. "The successful use of the Taguchi methods to increase manufacturing process capability." *Journal of Quality Engineering*, Vol. III, 1991, pp. 51–62.
Shewhart, W. *Economic Control of Quality of Manufactured Products.* New York: Van Nostrand Co, 1931.
Snedecor, G. and Cochran, W. *Statistical Methods*, 6th edn. Iowa State University Press, 1967.
Western Electric Company. *Statistical Quality Control Handbook.* Easton, Pa.: Mack Printing Company, 1956. (Original AT & T *Statistical Quality Control Handbook.*)

CHAPTER 5

Organizing, Managing, and Measuring Concurrent Engineering

This chapter will discuss nontechnical but very important aspects of concurrent engineering: organizing, managing, and measuring the interdisciplinary effort required. When introducing these concepts to an organization, it is important that the different parts of the organization see this effort as an opportunity to represent their particular needs and concerns; yet they have to agree on a common set of goals that benefits all. The results of not paying attention to team building and reducing interdisciplinary conflict could put the concurrent engineering effort at risk.

Concurrent engineering might run contrary to the historical methods by which design engineers' performance has been measured. One of the important aspects of developing new products is "time to market." Paying attention to concurrent engineering efforts, which might involve some training and organizing of multidisciplinary teams, might initially be considered as contrary to the goal of shortened time to market. Many design engineering managers have resisted concurrent engineering because of this aspect. What they eventually find is that the overall cycle is shorter because there are fewer revisions and design iterations due to problems encountered later in that cycle (such as manufacturability and serviceability issues).

The long-term benefits of concurrent engineering (CE) and the methods to institutionalize CE in the organization have to be quantified. This chapter will discuss the issues of building and organizing multidisciplinary teams and the measures of the concurrent engineering process: The elements of building an effective team, the stages that the team goes through before focusing on the tasks at hand, the resolution of conflict, and the role of the team leader will be presented.

104 ORGANIZING, MANAGING, AND MEASURING CONCURRENT ENGINEERING

An important part of the new product development team's success is the company's environment and progress towards improving the design process and the development of its staff of engineers. This chapter will present a set of measures to monitor progress in these areas. They also keep track of the design and production phases and help new product development teams focus on nonproduct specific goals. These measures are intended to quantify the design and manufacturing phases of new product development, as well as the design process and investments in developing the design engineers.

5.1 FUNCTIONAL ROLES IN CONCURRENT ENGINEERING: DESIGN, MANUFACTURING, MARKETING, QUALITY, AND SALES

An important step in implementing CE in an electronics company is the recognition of CE as an important part of the company's competitive strategy. CE should be included in the goals and objectives of the overall organization, with each department having its own set of strategic plans that match the overall CE plan.

The benefits of concurrent engineering are well documented, such as shorter development time, fast ramp-up of production, lower product cost and higher quality parallel the traditional goals of these departments. Instituting a concurrent engineering program requires a serious commitment from every level of management and a close look at the design, engineering, and manufacturing processes. Managing this change positively in the company's methods and procedures requires careful planning and facilitating to ensure success.

The ownership of concurrent engineering should not belong to one particular group but should be shared equally among all. The role of the company management is to understand some of the inherent process changes in concurrent engineering, such as longer initial development cycle but reduced overall cycle and the measurement and continuous improvement of the current levels of product cost, testability, quality, reliability, and serviceability. In addition, it is important for the management to understand the issues of concurrent engineering and to set operational goals and measures that are in line with current product design and development practices.

5.1.1 The Role of the Design Engineering Department in Concurrent Engineering

The design engineering department's role in concurrent engineering is the understanding of the issues of manufacturability, testability, reliability, and other "-ilities." New product plans should be developed only after clear understanding of the current levels of "-ilities" in existing products. In many companies, these levels are not apparent. It is important that the design department does not estimate these levels but insists on the appropriate department, whether it is manufacturing, quality, marketing, or sales development to supply them with their best estimates of these numbers in order to set baseline levels and ultimate goals for new products.

In establishing a new product development strategy, each of the concurrent engineering plans and goals should be clearly outlined. The goal statements and the action plans used to achieve them should be formulated with cooperation from the other departments. The milestones and major project checkpoints for new products should contain progress updates on concurrent engineering goals.

Another important role for design engineering is to influence the long-range capital and process developments in the company. The long-range plan of manufacturing for increased automation, computer-integrated manufacturing, and information flow should be in line with plans for new products and technologies. For example, investment into further automation and manufacturing cycle reduction for through-hole PCB technologies should be discouraged, as the electronics industry switches to surface-mounted technology (SMT) and tape automatic bonding (TAB). In addition, the design department should influence other departments to obtain the information necessary to implement successful and credible concurrent engineering plans, such as documenting process capability for the manufacturing process, determining the current level of warranty cost for the quality department, and planning the service level for future products.

After the new product is released to manufacturing, a retrospective analysis should be performed by the development team. It should take place some time between 6 and 12 months after release to production, and before the project team is disbanded. The retrospective analysis should compare the actual results of the product performance after release to production with the original development project goals. The reasons for successes and failures in meeting the different "-ilities" should be brainstormed, documented, and presented to the management. Successes should be institutionalized. Failures should be corrected and the information fed back to new projects.

5.1.2 The Role of the Manufacturing Engineering Department in Concurrent Engineering

Manufacturing engineering department's role in concurrent engineering is the characterization and documentation of the current process, and its communication to design engineering. Design guidelines for existing and future manufacturing processes should be published, and updated to the most current state as equipment is purchased and the processes are enhanced. In addition, manufacturing has to control and continuously improve the quality of the current process, and outline its long-range plans in terms of equipment, people, and information flow.

Characterizing the manufacturing capability and constraints is the key to the success of concurrent engineering. Process capability measurements can be a direct result of maintaining statistical quality control on the production process. A target plan for continuous process improvement and its results should be communicated regularly to the design engineering department. Failure data should be considered not only from factory processes but from field failures and warranty reports as well.

The long-range plans for manufacturing in terms of process capability, automation, test, supplier certification, delivery and distribution, and people training and recruiting efforts should be made in line with the company market strategy, and after consultation with the engineering department. Technology risks in manufacturing plans, such as the decision when to maintain an old technology or to jump ahead with a new technology, should be made in light of the new product development plans.

5.1.3 The Role of Other Departments in Concurrent Engineering

Other departments, such as quality, marketing, and field service, should also be involved in the concurrent engineering process to set the current baseline of quality, reliability, service, and repair of current products.

Marketing should play an important role in focusing customer inputs by using tools and techniques such as quality function deployment (QFD). QFD is a common marketing and design engineering tool. It is similar to other concurrent engineering tools, such as process capability index and design efficiency ratings, that bind the design and manufacturing parts of the organization together. These tools form a common language and partnership between the design department, marketing, and manufacturing in order to focus on the process rather than the merits of the design or the personalities involved.

The quality department should provide the audit function on the quality data being generated at the production floor, as well as the reliability data being generated in the field. Quality should not be the sole responsibility of the quality department but should be the concern of all parts of the company. In more recent times, quality departments have been active in introducing

DESIGN GUIDELINES 107

these tools to the overall organization, because of their traditional role as customer advocates.

The field service department should input very strongly into the design of new products in order to facilitate the serviceability and the repair of electronic products. Many of the tenets of concurrent engineering, such as standardization of parts and ease of assembly, should result in reduction of the number of tools and spare parts that the field support organization needs to stock. Concurrent engineering emphasis on eliminating fasteners, and increasing the use of snap fits, makes the disassembly of products for repairs more difficult. The field service department should participate in the design decision.

5.2 DESIGN GUIDELINES

A primary vehicle for communicating manufacturing capabilities and constraints is the publication of *design guidelines*. This document should be developed and updated on a regular basis and communicated to the design engineers. It is important that design guidelines should anticipate the release of new products 6–18 months into the future, and they should allow design engineers to include process improvements in their design specifications. A rigid set of design guidelines that is not in step with technological improvements will achieve the opposite result, by constricting available technological enhancements, and will lead to frustrations and conflict between the design and manufacturing engineers.

Manufacturing equipment suppliers have traditionally responded to customer's needs by issuing design guidelines for their equipment. An example is the design guidelines for electronic component insertion and placement published by Universal Instrument Corporation for automatic assembly equipment. These supplier guidelines can form the basis for in-house design guidelines, even if the specific equipment is not used in production.

In addition, societies such as the Institute of Printed Circuits (IPC) continually publish process and manufacturing guidelines for PCB design, fabrication and assembly, as well as general assembly requirements. These design guidelines are specific to a class of PCB technologies or capabilities. The specification of the IPC and other similar societies, such as the Society for Manufacturing Engineers (SME), should be used to form the basis of design guidelines and documentation for future manufacturing technologies.

Design guidelines should be treated as any other company specification, under engineering change control and management sign-off. In many cases, with the advent of computer-aided engineering (CAE) design workstations, the design guidelines could be imbedded in the CAE process configuration. Examples are PCB design guidelines that can be loaded into PCB Design software to identify PCB fabrication processes such as foil widths and spaces. In sheet metal design, standard punch patterns can be loaded by the CAD software from the part database and be available to design engineers. Design guideline methodology for assembly and sheet metal and plastics fabrication will be discussed in later chapters.

The design guidelines should be augmented with the manufacturing process future plans, the outline of the schedule of process and technology upgrades, and with the process improvements updates based on actual quality monitoring on the production floor. It is also useful to publish process cost estimates for typical operations in the plant, such as the cost of assembly and test, either manual or automatic, for electronic components. The defect rates should also be published for each process, so that the test and repair costs can be accurately estimated by the design engineers.

5.2.1 Design Guidelines: In-house Versus Supplier Capability

One of the issues of design guidelines is that of in-house versus outside suppliers' capabilities and constraints. This issue arises when the company's internal manufacturing process is a subset of external contractors' capability: An example is the image transfer of finer lines and spaces for PCBs available at contractors' facilities versus the company's captive PCB fabrication facilities.

This issue of in-house versus suppliers' manufacturing capability and constraints is a difficult one, fraught with emotions of company loyalty and the job security of the company's employees. The answer to this dilemma to force the internal company capabilities and constraints to keep up with outside technology in the long run. Otherwise, the competitive position of the company could be jeopardized. An outdated manufacturing facility, even if run with high quality and production efficiency, will invariably be more costly in the long run, and will have to either compete with the outside suppliers or shut down.

This philosophy of assuming the best current technology available, and not being restricted to the in-house capability, should be developed with full communication through the organization. Future equipment purchase plans, training schedules, and process enhancement targets should be focused on achieving common goals and staying technologically current within the company. If the company continues to pursue internal specialized processes, it may benefit in the short term, but will ultimately lag the industry in the long term. It is very difficult for a company to sustain special processes and keep them current in view of the rapid technological improvements in materials and technologies.

5.3 ORGANIZING FOR CONCURRENT ENGINEERING

The project team approach has been proven to be the most successful organizational structure used to implement new product development, because of its focus on the product. One of the important factors in the future success of the concurrent engineering effort is the acquired knowledge of "-ilities" in the design and development engineering team. Manufacturability or serviceability knowledge does not have to be represented on the project team by an engineer from that area, because of the extra cost involved, but should be common knowledge to all team members. The use of design guidelines and periodic design reviews for manufacturability or serviceability is one method of maintaining the focus on these issues. Educating the design engineers in these issues, either through rotating engineering assignments to other areas of the plant or actually having the other departments provide training and knowledge in this area, is also effective in increasing the total product and process knowledge of the design engineers.

This training phase of the team members will take some time away from the product development effort. New product project managers may be reluctant to let their engineers spend the time initially because of the concern for the development schedule. In most cases, the ultimate benefits of this initial training and concurrent engineering focus phase have resulted in an overall shorter project time because of the reduction in problems encountered later in the project.

5.3.1 Qualities of Modern Design Engineers

The approach to defining the qualities of modern design engineers in the arena of CE and DFM is not different from the traditional ways that engineers of superior performance (and hence those highest paid) were recognized. The most important qualities of the modern designers are their abilities to meet project goals on time, with low cost and high quality achieved. These top performers owe their success to their good technical knowledge of the design and manufacturing processes, as well as a high degree of creativity used in perceiving future problems and solving them promptly.

Recently, a new attribute of design engineers that is important to the success of concurrent engineering has been gaining and rates equally with technical knowledge and creativity—*teamwork*. With the focus of teamwork in interdisciplinary projects, this aspect has become critical to the success of the overall project. Teamwork is not a natural quality of engineers, since they tend to be competitive and are encouraged to become so by the grading systems at colleges and universities. Team building training and concepts are very important to the future success of concurrent engineering in electronics companies.

With the adoption of many of the techniques and methodologies of concurrent engineering, such as design guidelines, QFD, simulation, analysis tools, and design efficiency ratings, engineers are concerned about the loss of creativity and design freedom. They feel constrained by the fact that these methods require a specific path to achieve the design objectives. There is little room to maneuver and creating breakthroughs in the design process.

The creativity of design engineers can be released through implementation of the different steps of concurrent engineering methodologies, by taking the product concepts and process specifications and physically creating the design features of the products. The tools and techniques of concurrent engineering help designers meet the most important part of their job, which is completing the design on schedule. Once the elements of the design process have been outlined and understood by all, there will be very few short-cuts. In fact, engineers spend a small amount of their time doing design and analysis work; most of it is spend on communication, planning, and documentation. A typical result of a survey conducted at major companies in the process of justifying CAE/CAD equipment shows where design engineers spend their time in a nonautomated environment:

Design (thinking, theorizing) and analysis (modeling, simulation)	10%
Testing (data reduction, prototypes)	10%
Documenting (memos, drawings, sketches)	25%
Planning (Gantt/Pert, scheduling)	15%
Communicating (meetings, sharing information, talking, listening)	40%

Although the data is for a nonautomated (non CAE) environment, the fact is that CAE is only effective in reducing design, analysis, and documentation time, which is about a third of the total time spent at work. These results, which have been confirmed at many companies, indicate the necessity for teamwork and communication skills by the design engineers to complement their technical competence and creativity.

5.3.2 Building an Effective Product Development Team

An effective product development team is a synergistic group of engineers who are committed to achieving common objectives: working well together;

sharing resources, information, and skill sets; using and learning from collective experience; and producing high-quality results.

New product development teams will have to be formed with experts from different parts of the organization. In many cases they will be working together for the first time. To successfully accomplish the development projects, a team building process will have to take effect to ensure the project's success. The team building process includes deliberately creating a team to meet a certain set of objectives and goals. Each team member's roles and responsibilities have to be well defined, as well as the objectives and goals of the whole team. In addition, the criteria for measuring the team's progress and success factors has to be agreed upon in advance.

Effective teams have many similar characteristics that are essential to achieving a high degree of success:

- Meetings are conducted in a relaxed, informal and supportive manner. Team members generally enjoy working together, are involved in the task at hand, and are continuously challenged by the technical aspect of the project.
- The goals and objectives of the team are well defined and each team member is consulted and agrees with them. There is basic agreement among team members on working towards the same objectives, and each member knows each other's opinion about the issues being discussed.
- The skills sets that members bring are divergent, but each team member is credible in his or her knowledge and experience in their area of expertise.
- Every team member is respected for their knowledge in their own area, and team members listen and respect each other's ideas. The basic tenets of brainstorming are used freely in making decisions, where team members feel free to participate in discussions, make suggestions, express ideas and concerns without fear of premature judgment, conclusions, or ridicule.
- The team quickly arrives at a set of informal procedures to conduct team business. The team establishes a pattern for problem solving and goal setting, including problem statements, data collection and analysis, and effective decision making. All problem statements, decision analysis, and resolution are done collectively.
- During team meetings, discussions are animated, disagreements are common, and conflicts are dealt with directly, openly, and quickly. Techniques for problem solving are used effectively, focusing on the issues, without getting embroiled in personalities. Criticisms are constructive, and aimed at ideas not at people. The group focus is

working together toward removing obstacles that prevent further group progress.
- Decisions are mostly arrived at by consensus, with team members working together toward collective agreement. When some team members do not individually agree with the results, they will nevertheless have to support the group's decisions.

These team building and mode of operation concepts are important in achieving cohesive teams that are focused on the project at hand and that will have a fairly high probability of success. It is important that the team be well balanced, from grisly veterans of past product successes to new engineers who are on the upward swing of the learning curve. The achievement of personal success and professional advancement through team cooperation and strength is the formula for the total success of the organization.

5.3.3 Stages of Team Development

There are four stages in team development that are passed through from the time the team is formed to its maturity while getting focused on its tasks:

1. *Probing.* In this phase, the team members try to define their roles and positions relative to other group members. The group values and procedures are tested during this stage, before arriving at a group norm.
2. *Positioning.* In this stage, team members begin to forge alliances and specific team members establish themselves as significant and powerful, through knowledge, experience, or devotion to the task. During this period, the team members(s) who are in control of the team are established, as well as their methods of controlling the team. These may be by steering conversation, by acting as arbitrators in arguments, by providing outside information or getting things done through others. Nonconforming team members begin to toe the line, having had their position altered by the group norm.
3. *Organization.* This stage marks the beginning of taking on the task at hand. The team members establish their commitment in working together, and the team's objectives become clearly defined and supported by team members. The methodology of operation, data collection and analysis, problem solving, conflict resolution, and decision making are well established.
4. *Maturity.* During this stage, the team is performing its tasks, and is encouraged by accomplishing early milestones. There is an atmosphere of mutual respect, informality, cordiality, and rapport. Team members share a common view of the project's progress, and act as informal spokesmen for each other and the project.

ORGANIZING FOR CONCURRENT ENGINEERING

These stages have been shown to be repeated many times in team-forming situations. Teams that are continuously being changed, either through additions of other team members or transfers to other departments, lose their continuity of development and have to reestablish these stages as the team's make-up changes. It is important to size the project team correctly to the development task, as well as to define a good set of goals and to keep the team intact.

5.3.4 Resolving Team Conflict

Before the advent of concurrent engineering, "tossing the product over the wall to manufacturing" was the norm of introducing new products. With concurrent engineering, other departments participate through the new product development process, either through actual inclusion in the new product teams or through attending and contributing to design reviews and checkpoint meetings.

In these instances, conflicts about the new product's manufacturability, testability, reliability, serviceability, and other "ilities" will arise. The typical comment from the other departments is "This design is not acceptable." The typical designers' response is, "It's too late to change the design."

Concurrent engineering will not eliminate these conflicts. Hopefully it will bring them to light early enough in the design stage that either the new design initially meets the requirements, or it is very easy to change it to meet the "-ilities" without incurring large costs in the redesign of materials, parts, or processes.

Understanding dissension, its causes, and methods to resolve it are helpful to the project team. Conflicts arise when issues are brought up to the group. They can be substantive, such as value differences, divergent goals, differences in perceiving the problem, and confusion about assigned role. Or they can be personal, such as incompatible styles, personality traits (ego), or a desire for independence.

There are varied responses to conflicts depending on the personal versus group relations disposition of team members. Compromise is the middle ground. Team members who do not care about their relationship with the rest of the team take on the risky path of win or lose. Team members with a high regard for group relationship take on the path of least conflict, by either accommodating or collaborating.

The best conflict resolution strategy is the avoidance of the conflict before it starts, by addressing the issue at hand and using problem solving techniques: more data collection, problem analysis, and decision making. Negotiation and compromise are the key. Otherwise, the antagonists take on the dangerous path of a win/lose strategy of conflicts, with serious consequences for the success of the team.

5.3.5 Role of the Team Leader

The leader is the representative of the team to the management. His or her job is to act as the conduit by which the project team continues to receive proper resources of people, equipment, material, training, and support from the organization. He or she reports on the project schedule, and alerts management of upcoming concerns and difficulties. The successful project team leaders are not the ones who autocratically manage by fear, but rather those who command respect and confidence among team members.

An important aspect of the management role for the product development team leaders is the focus on "people issues." The team leader can best accomplish the team's mission as follows:

- Select and recruit a well balanced team, in terms of technical knowledge, skills and experience. A team should not be made totally of senior experienced engineers but should leave room for new engineers to grow while experienced engineers mentor them. Time and resources for team development should be made available: training, team building, and enhancing written and oral communications.
- Assign responsibilities that match team members' capabilities. Allow each member to reach their potential by giving them free rein in their areas of expertise. Evaluate each member's progress, solicit opinions from throughout the organization, and provide timely and correct feedback. Always give credit where credit is due.
- Keep the team well informed on the management perspective and the state of the project and the company in general. Facilitate and encourage communications and free flow of ideas, both within the team and with other departments. Allow team members to make decisions in their domain, and to represent the team in making presentations to management and negotiations with other departments and suppliers.
- Reorganize the team to meet changes in project goals and technical needs. Recognize when a stumbling block is preventing further progress. Seek help from the team members, management, and other sources.
- Act as the team's point person in recognizing and removing obstacles. Resolve conflicts by seeking problem solving techniques. Use directive management techniques when necessary, to quickly resolve a problem and get the project on track.

The project team leader is the most important member of the team. Project leaders are developed, not born. With the correct mix of leadership, delegation, and technical knowledge, project leaders will complete their projects successfully.

5.4 MEASURING CONCURRENT ENGINEERING

Measuring concurrent engineering involves measuring the aspects of the development process that are not directly related to product development and achieving product performance milestones. These measures should be kept up to date and used to set the goals of new product development projects. They are divided into four categories of performance metrics:

Design phase metrics
Production phase metrics
Design process metrics
People metrics

A summary of these metrics is presented in Table 5.1.

These metrics are intended as a starting kit towards identifying important aspects of the design and development process. Many of the measures can be set to historical levels, or updated as the company's competitive position changes. The design phase metrics are concerned with minimizing the level of engineering changes, and with factors applied to the definition to normalize these measures. They also measure the return factor (RF) of the project and the attention to staffing level and customer focus.

The production phase metrics are focused on the inherent benefits of concurrent engineering: ramp-up of production, minimizing the number of engineering changes after product release, and the cost and quality of the product, both at the factory and in the field. Again, normalizing factors are applied in the definitions of the measures to determine the effects of technology and complexity of the product evolution.

The design process metrics measure the investment in capital equipment and processes for the company, and keep track of the progress on turnaround times for prototypes and assemblies such as PCBs, plastics and sheet metal parts. In addition, the trends of these processes are also monitored to keep them within the general industry and competitive standards.

The people metrics measure the most important element of product development: the engineers and scientists working on the projects. Keeping the technical staff interested and motivated though prompt evaluations, training, and solid project assignment is very important to the long-term commitment of the company to its people.

TABLE 5.1. Performance Metrics for Measuring CE

Design Phase Metrics

	Measures	*Definition*
Project/Product		
Return factor	Return on investment	Incremental profit (5 years) / Project cost
Design engineering changes	Stability of design	Number of engineering changes after GO / Engineering change cost
Design iterations	Concurrent design	Number of iterations (PCB, mech) / Design complexity factors
Staff level	Staffing for success	Assigned staff / Required staff
Customer visits	Understand customer	Number of visits per engineer / Number of visits to different customers

Production Phase Metrics

	Measures	*Definition*
Product/Process		
Production readiness	Ramp-up to full production	Months of production / % of mature volume
Design quality	Engineering changes after production release	Number of engineering changes to assembly / Total number of released assemblies

Production quality	First-time yield	First-time yield / Mature product yield
Production cost	Labor standards	New product standards / Mature product standards
Design quality	Field failure rate	Failure rate / K$

Design Process Metrics

	Measures	Definition
Process		
Cycle time	Process turn around time	Time/process, PCB, sheet metal, etc.
Process trends	Long-term gain	Process parameters versus project time
Input/output	Process efficiency	R&D hours/process / R&D $/prcess
Productivity tools	Investment in tools	$ Investment in tools / $ R&D budget

People Metrics

	Measures	Definition
People		
Performance evaluation	People issues	% on time
Project completion	Job satisfaction	Number of projects completed / Number of projects started
Training	People investment	Number of hours/engineer
Turnover	Morale	% turnover/year

5.5 CONCLUSION

The organizational and team building aspects of concurrent engineering are just as important as the technical methods and techniques in developing successful new products. As companies introduce the concepts of concurrent engineering, this will involve different parts of the organization working together for the first time. Past conflicting departmental goals, missions, and adversarial relationships, if not managed correctly and positively, could adversely affect the success of the project team.

Teamwork, as a measure of engineering effectiveness, should be recognized as an important trait of the modern engineer, just as important as technical knowledge and creativity. It should be recognized in the engineer's performance evaluation and feedback. Creating a cooperative environment with emphasis on teamwork and reducing competition is an important step for the long-term health of the organization.

The role of each department in the organization towards instituting concurrent engineering as part of their goals setting process has been outlined. Methods of building a successful team, ways of reducing conflict, and the role of the team leader in achieving successful team efforts have been presented.

The measures of the R & D process metrics in terms of the progress in the design and manufacturing phases, as well as the improvements in the design process and the people metrics, are presented in a framework to encourage the concurrent engineering aspects of producing successful products with minimum development time and with high quality and low cost.

SUGGESTED READING

Eekels, J. "The decision structure in the management of multidisciplinary design projects." *Proceedings of the ICED*. Zurich: Heurista, 1985.

Deming, Edwards. *Out of the Crisis*. MIT Center for Advanced Engineering Studies, 1986.

Faraci, P. "New plotter design draws on teamwork." *Automation Magazine*, March, pp. 44–46, 1988.

Garret, Ronald. "Eight steps to simultaneous engineering." *Manufacturing Engineering* (Journal of the SME), November, 1990.

Gunn, Thomas G. *Manufacturing for Competitive Advantage*. Cambridge, Mass.: Ballinger Publishing Company, 1982.

Hall, Robert. *Zero Inventories*. Homewood, Ill.: Dow Jones Irwin, 1983.

Holusha, John. "Beating Japan at its own game." *The New York Times*, July 16, Section 3, pp. 3–7, 1989.

Hunsaker, P. and Alessandra, J. *The Art of Managing People*. Englewood Cliffs, N.J.: Prentice-Hall, 1980.

Imai, Masaaki. *Kaizen, The Key to Japan's Competitive Success*. New York: Random House Business Division, 1986.

SUGGESTED READING

Kerzner, Harold. *Project Management.* New York: Van Nostrand Reinhold, 1979.
Mohr, William. *Quality Circles.* Reading, Mass.: Addison Wesley, 1983.
Nikou, E. Communications plus team building = product success. *Printed Circuit Assembly Magazine,* April, 1989.
Shina, S. "The Technologist." *Keeping Pace with Change, The Challenge for Engineers.* A joint conference of Northeastern University College of Engineering in Collaboration with the Massachusetts High Technology Council, September 1984.
Spartz, D. *Management Vitality: The Team Approach.* Dearborn, Mich.: Society of Manufacturing Engineers Press, 1988.
Spitz, S.L. "There is more to DFM than technology." *Electronic Packaging and Production Magazine,* September, 1990, pp. 29–30.
Schonberger, Richard. *World Class Manufacturing.* New York: The Free Press, 1984.
Schonberger, Richard. *Japanese Manufacturing Techniques.* New York: The Free Press, 1982.
Stuelpnagel, Thomas. "Total quality management." *Journal of National Defense,* November 1988, pp. 57–62.
Young, John A. "Technology and competitiveness: A key to the economic future of the United States." *Science Magazine,* Vol. 241, 15 July, 1988.

CHAPTER 6

Robust Designs and Variability Reduction

The concepts of the robust design and variability reduction have recently been used to demonstrate some of the sources of unnecessary manufacturing and ownership costs for new products: repair and scrap in electronic manufacturing, and customer dissatisfaction with poorly performing products.

Repairs and scrap in manufacturing occur because production operations produce parts that are not within specifications, either because of tight tolerances or because of production process variability. Using robust designs, the need to have narrow specification limits is eliminated and the product will operate satisfactorily within a wide production process variability. A good indicator of this ratio of design specification to production variability is the process capability index, discussed in Chapter 4.

Robust design and variability reduction have been used widely in the Japanese and U.S. industries, as expressed by the recent popularity of the Taguchi method, developed by Dr. Genishi Taguchi, and application of algorithms of many different techniques of design of experiments, analysis of variance, and variability reduction techniques. Most of the applications of these methods have been made in the production of process development phases of new products, and not as replacements for the traditional design engineering methods such as analysis and verification.

This chapter will address the issues of using robust design methods in new product design engineering.

6.1 ON-LINE AND OFF-LINE QUALITY ENGINEERING

Reduction of production variability is a method used to increase the process capability index and reduce the cost of quality. Variability can be addressed by using a combination of two strategies:

On-line control, where the focus is to maintain the current production processes within a specified area of variability through control charts, optimal maintenance and calibration of production processes and equipment.

Off-line control, where a proactive effort is aimed at reducing the process variability or increasing design robustness through defect analysis and design of experiments.

On-line control methods have to be instituted before attempting off-line control projects. No amount of design of experiments and defect analysis can rectify a poor-quality operation that is out of control. In that case, the benefits of off-line control improvement can only be felt temporarily, before being negated by a manufacturing operation that is out of control and where the production parameters, materials, and processes are constantly changing.

The sources of defects, as outlined in Chapter 4, are due to the interaction between product specifications and process variability. This interaction originates from one of two sources: Either the process is not centered, when the process output mean does not equal the target value, or the product and process variability, as measured by the standard deviation of the manufactured product characteristics, is too high. Either one or a combination of both can influence the product defect level.

It is much easier to identify, collect data on, and rectify the first situation, a process mean not equal to the target. Incoming materials and equipment settings and performance can be measured, and if not equal to target can be corrected by strict adherence to specifications. Materials properties such as geometrical tolerances, density, tensile strength, or hardness can easily be measured and rectified by working with the production personnel and suppliers. Equipment parameters such as temperature, pressure, speeds and feeds, and motion accuracy can be measured by calibration gages against original purchase specifications and readjusted as necessary.

The calibration of production equipment is usually achieved by using an instrument that is inherently more accurate than the equipment to be calibrated. In addition, the instrument's accuracy has to be certified, through traceability to the National Bureau of Standards (NBS). It is common to use calibration equipment whose accuracy and resolution are at least ten times finer than that of the equipment being calibrated.

The maintenance of production variability and keeping of the product mean equal to the target are best accomplished by using control charts. The subject of control charts was covered in detail in Chapter 4. The two types of variable control charts cover the average of the process characteristic in the X-bar chart and the variability of the process characteristic in the R chart. Attribute charts do not easily make the distinction in the reject rate between the mean versus the deviation of the process, and therefore it is more difficult to ascertain the causes of the defects.

Production operations that are in good control generate a small number of defects, normally less than 100 PPM. The total rejects generated during a single production day is small, and therefore each defect can be analyzed to determine its cause. In many cases, the tools of continuous process improvements (CPI)—such as those presented in Chapter 4: brainstorming, cause and effect and pareto diagrams, data collection and sampling—can be used to determine the most probable cause for each defect. If a deviation of the production process is found to be the cause, the process can be adjusted accordingly.

Reducing the variability of the production process is more difficult and requires a thorough examination of the sources of variability. Some of these causes are uncontrolled factors or noise. They can be generated from the following:

- *External conditions*, imposed by the environment under which the product is manufactured or used, such as temperature, humidity, dirt, dust, shock, vibration, and human error. These conditions are beyond the control of the design and manufacturing process planners. They are difficult to predict, and it is expensive to design specific characteristics to satisfy all of the possible conditions under which the product is expected to operate.
- *Internal conditions*, under which the product is stored or used, such as friction, fatigue, creep, rust, corrosion, thermal stress. These conditions have to be specified corectly within the normal use of the product. However, many customers will overuse the product but will expect that the product will continue to operate even beyond its normal range. Therefore, the design has to be made more robust to ensure proper operations beyond advertised specifications.

Off-line quality engineering is focused on improving the robustness of the product functionality in external and internal conditions of operation. It seeks to determine the best set of process materials and parameters in order to ensure that the product characteristics average is equal to the specified nominal and that the variability of the product characteristics is as small as possible. A set of designed experiments can be performed to find such an optimum level of parameters.

6.2 ROBUST DESIGN TECHNIQUES

Robust design is a variant on the design of experiments (DOE) methods used to optimize products and processes for performance to specifications.

It is a mix of several tools that has been developed recently to optimize performance. It combines elements of brainstorming, design of experiments, orthogonal arrays, analysis of variance, and two new terms designated by Dr. Taguchi—loss function and signal-to-noise ratio.

The *loss function* is defined by Taguchi in his book *System of Experimental Design* as "The financial loss to society imparted by the product due to deviation of the product's functional characteristic from its desired target value." It is a negative definition of quality that totals up the quality loss after the product is shipped. This loss is not normally calculated by product designers, since much of the data is not available. The loss can be tangible, as in service and warranty costs that companies have to pay to repair the product. There are other costs that cannot be measured quantitatively: loss of market share, customer dissatisfaction, and lost future sales.

Signal-to-noise ratio is an indicator of variability and has no relationship to a similar term used in electrical engineering. Use of this ratio helps in designing the product or the process to be robust enough to work properly within the variations to which the product is exposed. These variations can be either internal or external noise applied to the product. Internal noise can occur from the regular use of the product through manufacturing or use; external noise involves factors that are not normally under the control of the manufacturer or the customer. Common examples of internal noise include that due to storage or normal use such as wear and friction losses of machinery, heat stress and relaxation, and rust and corrosion. External noise is imposed by the customers' use and is usually affected by characteristics of the environment in which the product is supposed to operate, such as humidity, temperature variation, or overstressing of the product beyond published specifications.

Japanese companies used the robust design and other similar methods to improve the quality of their products and achieve an internationally recognized high level of quality. When U.S. companies began to be overwhelmed by overseas competition, they discovered that their Japanese counterparts could achieve higher quality standards using these new methods, *even when they manufacture to original U.S. specifications and drawings.* The secret is variability reduction, by having the product built exactly to specification and the spread of the product variability very small around the mean. In statistical terms, the product mean is centered around the target and the standard deviation is very small.

Improving the process capability requires the concurrent efforts of both product and process designers. Product designers should increase the allowable tolerance to the maximum that will still permit the successful functioning of the product. Process designers should center the process to meet the specification target and minimize the variability of the process.

Since robust design uses statistical experimental methods to develop the best parameter settings for optimizing a process or a design, the requirements of in-depth technical knowledge of the basic science necessary for the design being optimized is not critical. In most cases, the engineers responsible for the process or design perform the necessary steps to conduct the experiments after taking some training in the techniques of robust design and design of experiments, perhaps with the assistance of a statistician.

6.2.1 Steps in Conducting a Robust Design Experiment

Conducting a robust design experiment involves using many of the tools of quality engineering that have been outlined in previous chapters. The success of this effort is dependent on selecting the proper team members, focusing on optimizing and measuring the proper characteristics, and following the guidelines of the methodology. Steps in performing a successful robust design experiment are as follows.

1. PROBLEM DEFINITION The first task in performing a robust design experiment is to outline the goals of the experiment and to define the characteristics of the process or the design to be optimized. While only one characteristic can be optimized at a time, many characteristics can be measured while performing the experiments, and analyzed separately. The final parameter selection can be a mix of the recommended parameter settings, depending on the compromise of the different objectives of each target characteristic level.

Creation of the boundary of the product or design to be optimized is important. The experiment should not be constrained to a small part of the design. On the other hand, the experiment should not be all-encompassing in attempting to optimize a wide span of product steps or processes.

2. TEAM CREATION AND DYNAMICS A project team should be selected to conduct the experimentation and analysis. The team should be composed of people knowledgeable in the product and process, and should solicit inputs from all people involved in the design to be optimized. It is not necessary for them to have in-depth technical understanding of the science of the problem, but the team members should have experience in similar or previous designs. Knowledge in statistical methods, and in particular robust design techniques should be available within the team, either through a statistician or through someone who has received training in robust designs.

3. PARAMETER SELECTION Robust design methods do not include screening experiments that narrow down the selection of the parameter settings that optimize the design. However, the team members might consider this project as an opportunity to try out as many possible parameters or different levels and combinations of the same parameters. In this case, screening experiments might be performed to narrow the parameter selection.

Brainstorming techniques should be used to select the number of parameters and the different levels selected for each parameter. The selection process should outline parameters that are as independent as possible from other parameters, and hence be additive in controlling the characteristic to be optimized. This is extremely important in reducing interactions of parameters that are difficult to quantify statistically.

An example of selecting independent parameters and reducing their interactions is the case of an infrared conveyorized oven for the reflowing of surface mount technology printed circuit boards. The reflow process is characterized by (1) the ramp-up of temperature to the solder melting stage, and (2) remaining at solder melting temperature for a predetermined time.

The oven temperature can be controlled by setting several heater zones on the top and bottom of the oven, as well as by varying the conveyor speed. Rather than choosing each temperature zone and the conveyor speed as the parameters, the proper choice would be the ramp-up temperature and the reflow time as the parameters to be optimized. The parameter levels should be achieved by actually experimenting with the temperature zones and conveyer speed.

Proper selection of the levels for each parameter used in the experiment is the next task to be performed. Levels that are either too close together or too far apart in value should not be selected, because they do not represent a continuous measure of the impact of the parameter on the measured characteristic. If the project team is confident in the adequacy of the current design level, then three levels could be chosen, with the current level being in the center of a 20% span represented by the other levels. If there is little confidence in the value of the parameter, two levels could be selected. They should be well within the operating range of the parameter that will achieve a working characteristic.

Parameters with multilevels of more than three should be avoided, because they significantly increase the complexity and number of experiments to be performed. They will also be unnecessary if the parameter of interest turns out not to be significant in determining the quality of the chosen characteristic. Detailed analysis could be performed on the most significant parameters after the robust experiments have been concluded. One exception to the avoidance of multilevel parameters is use of four levels when there is an interest in determining the impact of the absence of a parameter while still being able to measure three levels around the current value.

The number of parameters and their selected levels is also bound by the number of standard orthogonal arrays (see Design of Experiment next) available to conduct the experiment. Only certain combinations of parameters and their levels are available for performance of the experiment. Compromise might be necessary to achieve economy in the design of experiment by selecting a set of parameters and levels that is very close to one of the orthogonal arrays.

4. DESIGN OF EXPERIMENT Robust design experiments are conducted using orthogonal arrays (see Section 6.3.2). There are only a small number of these arrays, and their size increases geometrically.

The arrays are arranged in terms of the number of experiments, parameters, and levels. The L8 orthogonal array, using the notation 2^7 (see the list of array types that follows), is an eight-experiment array, having seven columns to be used as parameters, and each parameter is to be considered at two levels. The next two-level array is L16, or 2^{15}, with 16 experiments and 15 parameters at two levels.

Orthogonal array types (36 experiments maximum)

Array name	Number of parameters	Number of levels
L4	3	2
L8	7	2
L9	4	3
L12	11	2
L16	15	2
L18	8	7 Parameters at 3 levels, 1 Parameter at 2 levels
L27	13	3
L32	31	2
L36	23	12 Parameters at 3 levels, 11 Parameters at 2 levels

It is not necessary to choose all available parameters to be included in the experiment: An L16 experiment can be performed with 10 parameters; the other parameters (array columns) are left empty. This does jeopardize the utility of the experiment, since the analysis of the effect of the 10 parameters on the output characteristic is valid.

L12 (2^{11}), L18 (2^1, 3^7), and L36 (2^{11}, 3^{12}) are combination arrays that are unique because there is no interaction of parameters that could produce hidden miscalculations if they were not carefully selected. Interaction will be further explained later in this chapter.

The orthogonal arrays are used to design the experiment by assigning different levels to each parameter, which then form each line of the set of experiments to be conducted. The measurements of the characteristic to be optimized have to be repeated using various scenarios, depending on the noise considerations of the design:

- To simulate individual part-to-part changes. Repeating the experiment three or four times is sufficient to study this type of design variability.
- To simulate noise conditions, experiment repetitions are required for each set of noise parameters, depending on the levels of noise desired. For example, to study the variability effects of three noise parameters at two levels each, 6 repetitions (3 parameters at 2 levels) of the experiment set and measurements for the desired characteristics are required. To minimize the number of repetitions, an orthogonal array could be used to determine the minimum set of repetitions. In the example above, instead of using six experiments, only four experiments are sufficient though an L4 (2^3), which is a representation of four experiments and three noise parameters at two levels each. This use of orthogonal arrays for noise simulation is called the *outer array method*.
- To study the full effect of the design tolerance and variability reduction, the number of repetitions can be determined by an outer array with an equal number of parameters used in the experiment. Thus the experiments in an L36 design (2^{11}, 3^{12}) have to be repeated 36 times giving a total of $36 \times 36 = 1,296$ times to have a full tolerance analysis of the design. This is quite cumbersome, and often computer simulations can be used to perform the entire tolerance variability set that the robust design is trying to minimize. However, the entire set would amount to $2^{11} + 3^{12} = 2,048 + 531,144$ or $533,489$ repetitions of the design analysis, resulting in a lengthy computer analysis run.

5. ANALYSIS OF DATA Once the experiment has been performed, the data can be analyzed to determine the optimal settings of the parameters' levels for adjusting the mean of the design to target and for minimizing the design variability. In addition, the significance of each parameter's effect on the design characteristic can be calculated through the use of analysis of variance (ANOVA). Using ANOVA, the significance of each parameter on the performance of the design can be calculated. Important parameters can be set to the proper level, and least-significant parameters can be ignored.

The significance test is only a statistical determination of the probability that the effect of the particular parameter on the characteristic measurement is not due to chance alone. Although there are techniques for improving the significance levels of parameters, graphical presentation of the data is sufficient to determine the best parameter setting in order to adjust the design mean to target and reduce design variability.

6. PREDICTION AND CONFIRMATION OF EXPERIMENTS One of the unique elements of robust designs is the prediction of the characteristic value to be achieved, especially in performing experiments for increasing the quality of designs. In addition, a range of the predicted characteristic value can be calculated, which increases the confidence of achieving the target.

After performing the experiments and conducting the data analysis, a choice of parameter levels has to be made as a compromise between setting the design characteristic mean to the target value and reducing variability. A recommended parameter level might cause variability to be reduced significantly, yet at the same time the process mean could be shifted from target. Another case occurs when multiple characteristics are to be optimized using one experiment with many separate output measurements and data analysis. For example, a robust design experiment could be performed to design a new plastic material. The material will have several desired characteristics including cream time, gel time, core density, flow, and free rise density. The experiments will be designed using an orthogonal array that determines what ratios of raw materials are to be used. Measurement of all the desired characteristics will be performed, then the data will be analyzed to determine the best set of raw material ratios for each characteristic. A compromise of all recommended parameter levels will have to be made in order to achieve the best overall plastic product.

Once all the choices and predictions of the robust design experiment have been agreed upon, a confirmation run should be made before final adoption of the design decision, to verify the theoretical analysis. This confirmation will test out the entire robust design process before full implementation takes place. The design should continue to be monitored in production through statistical quality control methods for a minimum of six months before any further attempts are made to increase the robustness of the design by using another set of experiments.

6.3 ROBUST DESIGN TOOL SET

Robust design techniques use traditional as well as new concepts for optimizing designs by moving the product characteristics to target and reducing their variability. Some of these concepts are as follows.

6.3.1 Loss Function

The loss function is a quadratic expression estimating the cost of the average versus target and the variability of the product characteristic in terms of the monetary loss due to product failure in the eyes of the customer. Traditional

ROBUST DESIGN TOOL SET

methods have been to calculate this loss only when the product characteristics exceed customer specifications. The robust design method estimates the loss based on the current product characteristic distribution around its mean.

The loss function L indicates a monetary measure for the product characteristic mean versus its target value and the distribution around the mean. Generally it is expressed in terms of the cost of each failure divided by the square of the deviation from the mean at which the failure occurs. Normally this is expressed as

$$L(y) = \frac{A}{\Delta^2}(y-m)^2$$

where
L = loss function
y = design characteristic
m = target value
A = cost of repair or replacement of the product
Δ = functional limit of the product

Rewriting the formula by using the fact that $(y-m)^2$ is similar to the expression for mean square deviation or the variance for the product characteristics,

$$L = \frac{A}{\Delta^2}\text{ (mean square deviation)}$$

The loss formula can be translted into the more familiar statistical terms of output characteristic average $\bar{Y}$ and the standard deviation σ. In addition, this formula can be simplified when the goal of the robust design is to make the characteristic as small as possible (such as reducing defects to zero) or making them as large as possible (such as increasing the strength or power of electrical signals). A summary of these formulas is given in Table 6.1.

Example: Loss Function in a PCB Connector

An example of the quality problems that occur in the fabrication of printed circuit boards (PCBs) is the fit of the PCB connector into the product housing or "card cage." If the variability of the connector size is large, the fit is difficult or impossible to achieve and could result in scrapping the PCB.

Assume that the tolerance for acceptable fit is ± 6 mm, that the cost of rejecting the PCB at the fabrication shop is $100, and that the cost of rejection at the customer site after the PCB has been assembled is $500. A typical lot of 18 PCBs from the PCB fabricator was measured. The following shows the calculations of the loss function due to the variability of the connector and estimates the savings incurred by either adjusting the mean to target or reducing variability.

TABLE 6.1. Quality loss function

Customer tolerance	Loss formula
Nominal is best $M^{+\Delta_0}_{-\Delta_0}$	$L = \dfrac{A}{\Delta_0^2}[\sigma^2 + (\bar{Y} - M)^2]$
Small is best $0_0^{+\Delta}$	$L = \dfrac{A}{\Delta^2}(\sigma^2 + \bar{Y}^2)$
Large is best Δ or greater	$L = A\Delta^2 \dfrac{1}{\bar{Y}^2}\left(1 + 3\dfrac{\sigma^2}{\bar{Y}^2}\right)$

A = failure loss
$\bar{Y}$ = process average value
σ = process/product standard deviation with respect to Y
Δ = deviation at which product failure occurs
M = target value of Y
$L(Y)$ = loss in $ when quality characteristic is equal to Y

Assume actual deviations of a set of 18 PCBs at fabrication shop
0, 0, −3, 0, 0, 1, 0, −5, −2, −2, 3, −5, −1, 0, −4, 3, 0, 1

$$L = K(Y - M)^2$$
$$= K(\text{MSD}) = K \text{ (mean square deviation)}$$
$$\text{MSD} = \frac{1}{N}(Y_1^2 + Y_2^2 + Y_3^2 + \cdots + Y_N^2)$$
$$= \frac{1}{18}\{0^2 + 0^2 + \cdots + 1.0^2\} = 5.78 \text{ mils}$$

Hence,
$$L = \frac{A}{\Delta^2} \times \text{MSD} = \frac{500}{36} \times 5.78 = \$80.25 \text{ per PCB}$$

Assuming that the mean will be set to the target value using a future robust design experiment, the expected savings are:
$$L_{\text{NEW}} = K(\sigma) = K(\text{variance})$$
$$= \frac{A}{\Delta^2} \times (\text{variance}) = \frac{500}{36} \times 5.48 = \$76.11$$

Hence,

$$\text{Quality improvement} = 80.25 - 76.11 = \$4.14 \text{ per part}$$

The importance of the loss function is that it gives a monetary value to the state of the output of the process, both in terms of the process average not meeting the target and of the process deviation. In the example outlined above, the average for all 18 measurements, was -0.78 mm and the standard deviation was 2.34. Normally, the maximum loss function for an assembled PCB that is out of specification is set at $500, and if it is in specification there is no loss. The loss function assigns a value for this lot based on its variability, as shown above for this example at $80.25. Using the formula, the loss due to the process mean not being equal to target is calculated to be $4.14, while the loss due to variability around the mean is $76.11.

Adjusting the process mean to target is usually an easy task. It can be achieved by adjusting the equipment or checking material performance. Its effect on the loss function is only $4.14. The more difficult problem is reducing variability. This can be achieved by performing a robust design experiment.

6.3.2 Orthogonal Arrays

Table 6.2 is an example of an orthogonal array, L8. It has 8 experiment lines and 7 parameter columns (A through G), and each parameter column has two levels, 1 and 2, which means that both a high and a low value can be used for that parameter.

Orthogonal arrays possess several unique qualities. They can be performed at several levels, but the most popular ways are at either two or three levels. One of their most important attributes is that they are balanced: While the level is constant in a column, all of the other levels in the other columns will be rotated through their values. In the L8 example, experiments 1 through 4, while column A has level 1 the levels in columns B through G contain both levels 1 and 2.

By using this balance of the array, the effects of factors can be determined efficiently. N factors at two levels would normally require 2^N experiments. Using orthogonal arrays, we can perform maximizing experiments with only $N + 1$ experiments. In the L8 array example, we can perform all experiments for 7 factors in 8 experiments versus 2^7 (i.e. 128) experiments in the classical design of experiments. For an L9 array, the comparison is 9 experiments versus 81 (3^4) for traditional experimentation (see Table 6.3).

In the L9 orthogonal array, it takes 9 experiments to perform a maximization of four factors at three levels. The average of the results of the first three experiments, Y1, Y2, and Y3, is the average performance of the product or process due to selecting level 1 of parameter A, while the other

TABLE 6.2. L8 Orthogonal Array

	L8 (2^7)								Interaction table						
No.	1	2	3	4	5	6	7	No.	1	2	3	4	5	6	7
1	1	1	1	1	1	1	1	1		3	2	5	4	7	6
2	1	1	1	2	2	2	2	2			1	6	7	4	5
3	1	2	2	1	1	2	2	3				7	6	5	4
4	1	2	2	2	2	1	1	4					1	2	3
5	2	1	2	1	2	1	2	5						3	2
6	2	1	2	2	1	2	1	6							1
7	2	2	1	1	2	2	1	7							
8	2	2	1	2	1	1	2								

Linear graphs for L8 orthogonal array

● indicates main effect
────── indicates interaction between two dots
Numbers assigned to dots and lines indicate column assignments

TABLE 6.3. L9 (3^4) Orthogonal Array

Exp. No.	A	B	C	D	Result
1	1	1	1	1	Y1
2	1	2	2	2	Y2
3	1	3	3	3	Y3
4	2	1	2	3	Y4
5	2	2	3	1	Y5
6	2	3	1	2	Y6
7	3	1	3	2	Y7
8	3	2	1	3	Y8
9	3	3	2	1	Y9

●————————●
1   3, 4   2

Reprinted by permission of the American Supplier Institute, Dearborn, Michigan.

parameters negate themselves by averaging out their two levels. The average of Y2, Y5, and Y8 is the effect of selecting level 2 of factor B. In this manner, the average of all 12 possible combinations (factors A, B, C, and D and their levels 1, 2, and 3) is examined in terms of attaining the best result for the product or process specifications.

Example: Use of Orthogonal Arrays in Optimization of the Bonding Process

The process involved in this optimization consists of glueing parts for a high-technology application. An adhesive called RTV was selected as the bonding agent. The parameters for precleaning and temperature curing after RTV application were arbitrarily selected. The product was not performing in the field, as several parts separated during use. A robust experiment was designed to optimize the process for maximum peel strength. It was decided to measure the peel strength by a special tool that is commonly used to determine the maximum outside pressure necessary to cause the parts to separate.

Experimental setup. A decision was made to use three-level orthogonal arrays in order to maintain the current level as the middle level. The levels of the parameters were varied upward and downward in order to observe their effects. Four factors were considered in the L9 array:

- The cure temperature at 30°C (room temperature), 50°C (the current mild oven bake) and 70°C (a higher level of oven bake).
- The effect of ultrasonic cleaning, which was used in precleaning the parts, was to be tested after immersion for 1, 3, and 5 minutes, with 3 minutes being the current time.
- The volume of RTV dispensed, using different dispensing needs, was varied around the current volume of 1.7 ml at values of 1.2, 1.7 and 2.5 ml.
- The soak chemical used in the ultrasonic bath was varied from the current methylene (MT) to other cleaners such as methyl ethyl ketone (MK) and plain water (H_2O).

The experiment was designed as shown in Box 6.1. There were nine experiments, in which each experiment was a unique combination of parameter levels selected prior to its running. For example, in experiment number 3, 30°C was the cure temperature in the oven, the RTV volume was 2.5 ml, the ultrasonic cleaner was MK, and the cleaning time in the ultrasonic bath was 5 minutes. The result was a part that required 22.6 pounds force before it separated into two halves.

Data analysis. In this experiment, the data analysis was very simple and only required the use of a four-function calculator. The effects of each level were added, then tabulated as follows.

The highest average output for each parameter level can be selected, so that the ultimate bond strength can be maximized. In this case this is A3 (70°C), B2 (3 minutes), C3 (2.5 ml) and D1 (water for the ultrasonic bath).

The following interesting conclusions can be drawn: (1) The chart can be filled easily. Mathematical mistakes in the analysis table are eliminated, because all averages must add up to the same number (217 in this case). (2) The most important factor is the cure temperature (parameter A), since it causes the most change in

> **BOX 6.1: BONDING PROCESS OPTIMIZATION**
>
> In the bonding of chip transducers, several parameters were identified as important to the bonding strength:
>
> Parameters considered:
> A = cure temperature: 30   50   70   °C
> B = ultrasonic cleaning: 1   3   5   minutes
> C = RTV volume: 1.2 1.7 2.5 ml
> D = soak chemical: H_2O MT   MK
>
> **L9(3^4) Orthogonal Array**
>
> | Exp no. | A | B | C | D | Peel force |
> |---|---|---|---|---|---|
> | 1 | 30 | 1 | 1.2 | H_2O | 11.5 |
> | 2 | 30 | 3 | 1.7 | MT | 22.7 |
> | 3 | 30 | 5 | 2.5 | MK | 22.6 |
> | 4 | 50 | 1 | 1.7 | MK | 19.0 |
> | 5 | 50 | 3 | 2.5 | H_2O | 28.5 |
> | 6 | 50 | 5 | 1.2 | MT | 24.0 |
> | 7 | 70 | 1 | 2.5 | MT | 25.1 |
> | 8 | 70 | 3 | 1.2 | MK | 30.3 |
> | 9 | 70 | 5 | 1.7 | H_2O | 33.3 |
>
> A_1 = result of experiments 1 + 2 + 3
>   = 11.5 + 22.7 + 22.6 = 56.8
>
> **Average Outputs for Each Parameter/Level**
>
> | | A | B | C | D |
> |---|---|---|---|---|
> | Level 1 | 56.8 | 55.6 | 65.8 | 73.3 |
> | Level 2 | 71.5 | 81.5 | 75.0 | 71.8 |
> | Level 3 | 88.7 | 79.9 | 76.2 | 71.9 |
> | Total | 217.0 | 217.0 | 217.0 | 217.0 |

output. (3) The least important factor is the soak chemical (parameter D), since it hardly made a difference. (4) The average strength obtained by using the factors selected can be estimated by adding up the averages of the four factors (A3 + B2 + C3 + D1): 88.7 + 81.5 + 76.2 + 73.3 = 319.7/9 = 35.5. The expected strength from choosing the optimum set of parameters is A3 + B2 + C3 + D1. This

ROBUST DESIGN TOOL SET

represents approximately a 50% increase over the average of all experiments ($217/9 = 24.11$). (5) The combination of the best selected levels for the four factors (A3 + B2 + C3 + D1) is not within the L9 array table. By using robust design, it was only necessary to perform 9 experiments instead of 81 in order to find the optimum set of parameters levels.

Conclusions. As can be seen by the simple example above, robust design of experiments can optimize a process or a product easily and quickly by using very simple statistical techniques. It is also not necessary to have a deep understanding of the physics or the chemistry of the process or product to be optimized.

This particular example illustrates how the process mean can be shifted to the desired level (in this case to the maximum possible). A similar process can be applied to minimize the variability, or what is referred to as the signal-to-noise ratio. In this case, several products have to be run for each experiment line (four is preferable), and then the results can be maximized for reducing the variance as opposed to the average value as was done in the case study.

6.3.3 Signal-to-Noise Ratio

This expression is a measure of the variability of the output characteristic of the robust design experiments. The object of the signal-to-noise (or S/N) ratio is to express as a single number the repeatability of the output characteristic. This number can then be treated in a similar manner as in the analysis for improving the mean as was done in the peel strength example.

In the peel strength example, repeating the experiments four times would produce four sets of results for each line of experiments. The S/N ratio for each experiment line would be calculated from the formula for S/N ratio for the four repetitions. The formulas for signal-to-noise ratio are shown in Box 6.2.

The number of repetitions is dependent on the external conditions to be simulated by the robust design of experiment. Unit-to-unit variability can be simulated by several repetitions of the experiment. Specific noise conditions and their levels will determine the number of repetitions to be performed. Three noise factors, with two levels each, will require six repetitions of the experiments. As was indicated in the robust design methodology, an orthogonal array can be used in the outer array to reduce the number of repetitions—for example, from six to four using an L4 array.

In the case of the tolerance analysis of all the parameters, the repetitions will be calculated for all the experiment lines equal to the number of experiments. This technique is rather lengthy, as it will increase the repetition of an L36 array to 36×36 trials, or 1296.

> **BOX 6.2: ROBUST DESIGN STATISTICS**
> *Signal-to-Noise (S/N) Ratio*
>
> - S/N ratio is a measure of the uncontrollable, or noise factors.
>
> - Signal-to-noise analysis is similar to that for the mean except that different formulae are used for the experiment target:
>
> Nominal: $\quad 10\log_{10}\dfrac{\bar{Y}^2}{\sigma^2}, \quad \bar{Y}=\dfrac{1}{n}\sum_{i=1}^{n}y_i, \quad \sigma^2=\dfrac{1}{n-1}\sum_{i=1}^{n}(y_i-\bar{Y})^2$
>
> Smaller is better: $\quad -10\log_{10}\left(\dfrac{1}{n}\sum_{i=1}^{n}y_i^2\right)$
>
> Bigger is better: $\quad -10\log_{10}\left(\dfrac{1}{n}\sum_{i=1}^{n}\dfrac{1}{y_i^2}\right)$
>
> If the standard deviation for nominal is best stays the same regardless of the mean, use the formula for smaller is better
>
> - Do not confuse this signal to noise ratio with that defined in electrical engineering:
>
> $$\text{Signal-to-noise ratio}=\dfrac{\text{power of signal}}{\text{power of noise}}=\dfrac{\text{power of signal}}{\text{power of error}}$$
>
> Control factors = Assign to inner array
>
> Noise factors = Assign to outer array
>
> Error factors = Are included in S/N calculations
>
> - ANOVA should be performed for significant parameters.
> - A table should be made for tradeoff of factor levels that influence both mean and signal-to-noise ratio.

6.3.4 Parameter Interaction in Orthogonal Arrays

Interaction occurs when one parameter modifies the conditions of another. If this is deemed important, the interaction should be given its own column, but only at specific locations, where the interaction is demonstrated by the exclusive or (XOR) relationship. No parameter should be assigned to this column. See Box 6.3 for an example of interaction of two parameters.

ROBUST DESIGN TOOL SET

BOX 6.3: INTERACTION EXAMPLE

	No Interaction				*Interaction*		
	A1	A2	Total		A1	A2	Total
B1	74	78	152	B1	75	77	152
B2	80	84	164	B2	79	85	164
Total	154	162	316	Total	154	162	316

Plot of two mean values indicates the existence of interactions between them. Parallel lines = no interaction:

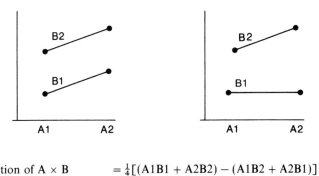

Interaction of A × B $= \frac{1}{4}[(A1B1 + A2B2) - (A1B2 + A2B1)]$

Mathematically, some parameters form an XOR relationship between their column assignments in the orthogonal arrays. XOR is a logical relationship with the following properties, assuming A and B as inputs, and C as an output:

XOR		
A	B	C
1	1	1
1	2	2
2	1	2
2	2	1

Array L8, columns 1, 2, and 3 form such an XOR relationship. These relationships are then grouped into primary parameters and interaction parameters. For example, columns 1, 2, 4, and 7 are primary parameters in array L8, while the remaining parameters are due to the interactions of these primary ones. Two possible scenarios for assigning interaction are as follows:

Scenario I			Scenario II	
Primary	Interaction		Primary	Interaction
1 × 2	3		1 × 2	3
1 × 4	5	or	1 × 4	5
2 × 4	6		1 × 7	6
7	none			

These relations are shown graphically through linear diagrams, which are shown with orthogonal arrays. See Table 6.2 and 6.3 for the linear graphs of L8 and L9. Thus all the parameters and their interaction are apportioned in one of two ways. As the array size increases, the number of choices in the parameter selections increases. These choices are expressed as linear graphs, which are graphical presentations of the parameter selection choices.

The choice of linear graph relationship depends on the robust experiments team's visualization of the design to be optimized. Scenario I presents an equal relationship between and importance of the first three primary parameters, with the last parameter considered as independent of all others. Scenario II assumes that primary parameter 1 is the most important, with all other primary parameters interacting with it. Whichever of the two scenarios the team selects, the experiment data analysis proceeds on that assumption, and the other scenario results cannot be calculated. The robust design team should be very careful when selecting which interaction scenario to use, and should spend an adequate amount of time brainstorming this issue.

Interactions have caused much confusion for robust design teams. If interaction is to be considered, only primary parameters can be used, which reduces the utility and economy of orthogonal array. For example, L27, which has 13 parameters at three levels, could drop down to only 7 primary parameters and 6 interactions. If more than seven parameters are to be used, then they should each be assigned to a column where the interaction is considered to be insignificant.

Interaction represents a mathematical indication of the effect of one parameter's variation on other parameters. If the interaction colum in the array is assigned to a parameter, then the analysis for the effect of that parameter could be either negated or amplified by the interaction effect. This effect is called *confounding*. In actuality, the effect of interactions is usually much smaller than expected. If interactions are of concern to the design team, then noninteracting orthogonal arrays such as L12, L18, or L36 could be used.

6.3.5 Analysis of Data with Interactions, Electrical Noise Example

In order to optimize the effect of electrical noise, an L8 experiment was performed with only four parameters being considered. The remaining three factors allowed in the L8 experiment were not assigned to parameters but were used to measure the interaction of the selected parameters. The selection of the parameter, and the graphical presentation of the mean values of each parameter level is shown in Box 6.4.

BOX 6.4: ELECTRICAL NOISE EXAMPLE

In reducing electrical noise, four parameters were considered, each with two levels. An L8 array was chosen.

1 = Cable connector type:	A	C
2 = Different contact methods:	spring	screw
4 = Conductive paint for the frame:	yes	no
7 = Number of shims:	0	2

L8 (2^7) Orthogonal Array

No.	1	2	4	7	Cable type	Contact method	Paint	Number of shims	Kilovolts
1	1	1	1	1	A	spring	yes	0	18.5
2	1	1	2	2	A	spring	no	2	14
3	1	2	1	2	A	screw	yes	2	18.5
4	1	2	2	1	A	screw	no	0	12.5
5	2	1	1	2	C	spring	yes	2	18.5
6	2	1	2	1	C	spring	no	0	13
7	2	2	1	1	C	screw	yes	0	9.5
8	2	2	2	2	C	screw	no	2	8

Plot the main effects:

Yes = 16.25
Spring = 16.00
A = 15.88
2 = 14.75
All = 14.06
0 = 13.38
C = 12.25
Screw = 12.13
No = 11.88

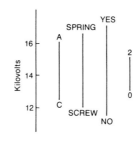

BOX 6.5: ELECTRICAL NOISE EXAMPLE: INTERACTION

In reducing electrical noise, four factors were considered: An L8 orthogonal array was chosen because it is the next highest array. Two interactions are free by good choice of factor columns

		A	C
A = Cable connector type:			
B = Different contact methods:		spring	screw
C = Interaction A × B			
D = Conductive paint for the frame		yes	no
E = Interaction A × D			
G = Number of shims		0	2

No.	A	B	A × B	D	A × D	G	Kilovolts
1	1	1	1	1	1	1	18.5
2	1	1	1	2	2	2	14
3	1	2	2	1	1	2	18.5
4	1	2	2	2	2	1	12.5
5	2	1	2	1	2	2	18.5
6	2	1	2	2	1	1	13
7	2	2	1	1	2	1	9.5
8	2	2	1	2	1	·2	8

Plot the main effects:

Yes = 16.25
Spring = 16.00
A = 15.88
A × B2 = 15.63
2 = 14.75
A × D1 = 14.50
All = 14.06
A × D2 = 13.63
0 = 13.38
A × B1 = 12.50
C = 12.25
Screw = 12.13
No = 11.88

- Note that the factors are arranged from most significant to least significant: Paint, Contact, Connector, Paint × Contact, Number of shims, Paint × Number of shims
- To decide which factors are significant use ANOVA Analysis
- To measure interaction of connector × Contact calculate the following averages:

A + spring,	A + screw,	C + spring,	C + screw
16.25	15.50	15.75	8.75

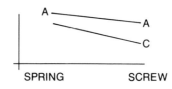

The interaction columns were those that were left empty. An analysis of those interactions was calculated as shown in Box 6.5.

6.3.6 Statistical Analysis of Robust Design Experiments

Analysis of the results of the robust design experiment is based on the analysis of variance (ANOVA), which is a method of determining the significance of

---

BOX 6.6: STATISTICAL DEFINITIONS

Target value Y_0
- The standard the experiment is designed to reach
- Could be theoretical, optimum conditions, regulatory, or based on competition's values

Sum of squares S_T
- Total sum of the squares of the deviations from target value. Deviation = $Y_N - Y_0$
- $S_T = (Y_1 - Y_0)^2 + (Y_2 - Y_0)^2 + (Y_3 - Y_0)^2 + \cdots + (Y_N - Y_0)^2$

When there is no target value, use correction factor, CF:
- $CF = (Y_1 + Y_2 + \cdots + Y_n)^2 / N$
- $S_T = Y_1^2 + Y_2^2 + \cdots + Y_n^2 - CF$

Sum of squares of parameter
- Square of sum of each level divided by number of levels − correction factor
- $S_p = (Y_1 + Y_2 + Y_3 + \cdots + Y_N)^2 / N - CF$

Sum of squares of the error, S_{ee}
- Sum of squares minus sum of squares of parameters
- $S_{ee} = S_T + S_p$

Degrees of freedom (DOF), f
- Independent squared items of associated statistics
- $f_T = n, f_L = $ Levels $- 1, f_m = 1, f_e = f_T - f_A - f_B \cdots$
  T is total number of parameters
  A is a parameter
  e is the error
- OA DOF = Number of parameters − 1

Variance, V
- Sometimes called MSD (mean square deviation)
- $V = S_T / f$

F-ratio
- $F = V_P / V_e$
  $V_P = $ variance of the parameter
  $V_e = $ variance of the error

each factor in terms of its effects on the output characteristic. The ANOVA analysis apportions the total effect of the output characteristic to each parameter in the orthogonal array. The significance test is based on the F-distribution, which is a ratio of the number of degrees of freedom for the parameter divided by the number of degrees of freedom for the error. The least-significant parameters are lumped together as the error of the experiment, since they are not important in affecting the output characteristic. More detailed information is required for determining the ANOVA analysis, which is beyond the scope of this chapter.

BOX 6.7: ELECTRICAL NOISE: ANOVA ANALYSIS

$CF = (18.5 + 14 + 18.5 + 12.5 + 18.5 + 13 + 9.5 + 8)^2/8 = 1582.03$
$S_A = [(18.5 + 14 + 18.5 + 12.5)^2/4] + [(18.5 + 13 + 9.5 + 8)^2/4] - 1582.0$
$\quad\quad = 26.28$

No.	A	B	A × B	D	A × D	G	Kilovolts
1	1	1	1	1	1	1	18.5
2	1	1	1	2	2	2	14
3	1	2	2	1	1	2	18.5
4	1	2	2	2	2	1	12.5
5	2	1	2	1	2	2	18.5
6	2	1	2	2	1	1	13
7	2	2	1	1	2	1	9.5
8	2	2	1	2	1	2	8

If factor has two levels, S_p is also $= (F_1 - F_2)^2/N_f$

$A1 = 18.5 + 14 + 18.5 + 12.5 = 63.5$
$A2 = 18.5 + 13 + 9.5 + 8 = 49$
$S_A = (63.5 - 49)^2/8 = 26.28$

Factor	DOF	S_T	V	F	
A	1	26.28	26.28	33.27	
B	1	30.03	30.03	38.01	No factors is
A × B	1	19.53	19.53	24.72	better than 95%
D	1	38.28	38.28	48.45	confidence level
A × D	1	1.53	1.53	1.94	
G	1	3.78	3.78	4.78	
Error, e	1	0.79	0.79		
Total	7	120.22	17.17		

BOX 6.8: ANOVA ANALYSIS: POOLED ERROR

When some factors are insignificant; they are combined with the error to obtain the pooled error, and ANOVA is repeated. In the example, G and A × D are pooled with error

The factors confidence are now:
  A 95%,   B 95%,   A × B 90%,   D 97.5%

Factor	DOF	S_T	V	F	S'	p
A	1	26.28	26.28	12.95	24.25	20.2%
B	1	30.03	30.03	14.79	28.00	23.3%
A × B	1	19.53	19.53	9.62	17.5	14.6%
D	1	38.28	38.28	18.86	36.25	30.1%
Pooled e	3	6.10	2.03		14.21	11.8%
Total	7	120.22	116.15		120.22	100.0%

The main effects will be improved by setting factors A, B, and D for bigger is better. Average of all factors and experiments = 14.06

Average of Significant Factors

A1 (connector A) 15.87   A2 (connector C) 12.25
B1 (spring)   16.00   B2 (screw)   12.13
D1 (paint yes)   16.25   D2 (paint no)   11.88

Choose: connector A; spring contact, paint

$$\text{Expected mean} = \text{average} + \tfrac{1}{2}(A1 + A2 + \cdots - A2 - B2 - \cdots)$$
$$= 14.06 + \tfrac{1}{2}(15.87 + 16.00 + 16.25 - 12.25 - 12.13 - 11.88)$$
$$= 19.99$$

Using the electrical noise problem, the analysis is performed using several of the terms for ANOVA. These are given in Box 6.6. Using these definitions for the statistical analysis, the calculations are as shown in Boxes 6.7 and 6.8.

6.4 USE OF ROBUST METHODS IN ENGINEERING DESIGN PROJECTS

The design engineering community feels that robust design is a technique to be used mainly for improving the quality of manufacturing processes. For design engineering tasks, the traditional methods of computer simulation and analysis of worst-case design studies of parts and materials remain dominant. Worst-case study is the method by which the engineer analyzes the design using a combination of the worst cases of the individual parts or materials specification limits.

A beneficial use of robust design methods is in performing robust design experiments in computer simulation of the design to obtain the experimental results. Thus an ideal design can quickly be identified from performance of an orthogonal array experiment input into the simulation. The results can then be analyzed for the optimal design.

Another use of robust design in new product development is in solving some of the "black magic" problems specific to electronic products. Examples are electrical noise reduction and reduction of radiofrequency (RF) interference to the level necessary to meet government regulations. Examples of parameters to be used include different cable routings, connector types, RF shielding bars, and cover types.

6.4.1 Comparison of Robust and Classical Design of Experiments

Classical design of experiments was historically used to arrive at the optimum settings of a process or a product design characteristic through a two step process:

Screening, in which a large number of possible factors are varied to determine which ones are to be included in further study.

Modeling, in which the factors found to be important in screening are further refined by experimental techniques.

When conducting classical design of experiments, the main concern is to find the optimum settings for the *mean* values of the factors to achieve a given characteristic. The purpose is to derive a theoretical framework for the linear relationship of the factors to the characteristic by regression analysis, where the fitness of the mathematical model to the theoretical performance of the system is measured by the regression factor R. A value of R higher than 0.90 is considered acceptable. Otherwise, factor selection must be reconsidered to achieve the proper model.

Another important difference is that the classical design of experiments is concerned with finding the optimum mean value of the parameter, while robust design has always been concerned about both the mean and the variability of the design.

6.5 CONCLUSION

It has been shown through several examples that robust design is an excellent tool for optimizing designs by shifting the mean characteristic of the design to target and reducing variability. The mathematical background for robust design is a mix of tools in orthogonal arrays, designed experiments, and analysis of variance. Several techniques in robust design should be thought out well in advance: the definition of the characteristics to be optimized, the selection of parameters and levels, the treatment of parameter interactions, and the simulation of variability.

The first robust design project should be selected carefully to optimize a design that is relevant but not too difficult in terms of complexity. Careful hand calculations should be made to complete the analysis. Only after initial successes should software-based methods of analysis be attempted.

SUGGESTED READING

Box, G. *Studies in Quality Improvement: Signal to Noise Ratio, Performance Criteria and Transformation.* Report No. 26, Center for Quality and Productivity Improvement, University of Wisconsin, 1987.

Cochran, W. and Cox, G. *Experimental Designs,* 2d edn. New York: Wiley, 1981.

Diamond, W.J. *Practical Experiment Design.* New York: Van Nostrand Reinhold, 1981.

Douglas, C. "Experimental design and product and process development." *Manufacturing Engineering Journal,* September 1988.

Ealy, L. "Taguchi basics." *Quality Journal,* pp. 26–30, Nov. 1988.

Guenther, W. *Concepts of Statistical Interference.* New York: McGraw-Hill, 1973.

Hicks, C. *Fundamental Concepts in the Design of Experiments.* New York: McGraw-Hill, 1964.

Holusha, John. "Improving quality the Japanese way." *The New York Times,* July 20, p. D7, 1988.

John, P. *Statistical Analysis and Design of Experiments.* New York: The MacMillan Company, 1971.

Kackar, R. and Shoemaker, A. "Robust design: A cost effective method for improving manufacturing processes." *AT&T Technical Journal,* 1985.

Lipson, C. and Sheth, N. *Statistical Design and Analysis of Engineering Experiments.* New York: McGraw-Hill, 1973.

Lowell, C., Shina, S. and Wu, J. "Optimization the new HOLLIS wave solder machine." American Supplier Institute 7th Symposium, Phoenix, Ariz., October 1989, pp. 101–116.

Ross, Philip. *Taguchi Techniques for Quality Engineering.* New York: McGraw-Hill, 1987.

Roy, Ranjit. *A Primer on the Taguchi Method.* New York: Van Nostrand Reinhold, 1990.

Phadke, Madhav. *Quality Engineering Using Robust Design.* Englewood Cliffs, N.J.: Prentice-Hall, 1989.

Shina, S. "Reducing solder wave defects using the Taguchi method." American Supplier Institute 6th Symposium, Dearborn, Mich., October 1988, pp. 123–144.

Shina, S. and Capulli, K. "Alternatives for cleaning hybrid integrated circuits using Taguchi methods." NEPCON EAST Conference, Boston, Mass., June 1990, pp. 931–942.

Stewart, C. "A time and a place for Taguchi." *Machine and Tool Blue Book*. December 1988, pp. 32–36.

Taguchi, Genichi. *System of Experimental Design*. White Plaines, New York: UNIPUB–Kraus International Publications, 1976.

Taguchi, Genichi. *Introduction to Quality Engineering*. White Plaines, New York: UNIPUB–Kraus International Publications, 1986.

Taguchi, Genichi and Konishi, S. *Orthogonal Arrays and Linear Graphs*, Foreword by Yuin Wu and Shin Taguchi. Dearborn, Mich.: ASI Press, 1987.

Tagushi, G., El Sayed, E. and Hsiang, T. *Quality Engineering in Production Systems*. New York: McGraw-Hill, 1988.

The Asahi, A Japanese Language Daily, April 15, 1979.

Young, Hugh D. *Statistical Treatment of Experimental Data*. New York: McGraw-Hill, 1962.

CHAPTER 7

Customer-Driven Engineering; Quality Function Deployment

7.1 INTRODUCTION

Quality function deployment is a structured and disciplined process that provides a means for identifying and carrying the customer's voice through each stage of product and service development and implementation. Quality Function Deployment (QFD) is achieved by cross-functional teams who collect, interpret, document, and prioritize customer requirements to identify bottlenecks and/or breakthrough opportunities.

QFD involves the functions of marketing, research, design, manufacture, quality, purchasing, sales, and service as required for a specific project. The QFD discipline provides both a framework and a structure for enhancing an organization's planning ability for communication, documentation, analysis, and prioritization. QFD utilizes a series of charts or matrices to help document the process for identifying breakthrough areas that will meet or exceed customer requirements. It works best within a company when there is organizational commitment and a disciplined approach to implementation.

QFD is a naturally compatible adjunct to concurrent engineering because it replaces the independent department responsibilities with interdependency and teamwork. It helps to tear down the barriers between manufacturing, design, research, and marketing. The design of new products is transformed from a separated series of steps into a process based on the voice of the customer that is integrated from design to production.

7.2 QUALITY FUNCTION DEPLOYMENT

7.2.1 History

Quality Function Deployment gained recognition in Japan in 1972 when the Kobe Shipyard of Mitsubishi used a matrixing technique to improve its design process. Its usage spread over the next few years, with many Japanese companies becoming disciples and heavy users of the QFD process. A few examples are Isuzu, Matsushita, Komatsu, NEC Micon.

In 1983 Maasaki Imai conducted a seminar in Chicago and introduced the participants to QFD, but little or no activity resulted from this event. In 1984 Don Clausing of MIT returned from Japan and introduced a basic model to the automotive industry. Ford Motor Co. became interested and John McHugh and Larry Sullivan of American Supplier Institute began working with Ford in initial QFD projects. Bob King of GOAL/QPC in his book, *Better Designs in Half the Time*, published in 1987, introduced a detailed model of QFD based upon the work of Dr. Yoji Akao of Tamagawa University, the leading voice of QFD in Japan.

Since 1987 QFD has become a key design and development planning tool for many companies in the United States. Some examples of companies and their applications include:

Clarification of engineering requirements—Ford Light Truck
Improved sales—Procter & Gamble hotel products
Improved internal customer/supplier relationship—Digital Equipment Corporation
Improved external customers/supplier relationship—Ford Climate Control, Cirtek, General Electric, and others
Improved manufacturing documentation and control—General Electric Motor
Improved hardware and software design—Hewlett-Packard and Digital Equipment Corporation
Improved education of new engineers—Cirtek
Prioritization and scheduling of design effort—Cirtek
Improved new product design and launch—Masland and Deere and Company
New design system—The Kendall Company
Clarification and prioritization of customer demands—Digital Equipment Corporation
Better understanding and documentation of customer demands
Established applicability in aerospace/defense electronics—Hughes Aircraft
Understanding who the customers are—Polaroid
Multicompany new product development—Rockwell International

7.2.2 Definition

Quality function deployment is an organized, disciplined process for determining the product or service requirements necessary to meet the stated

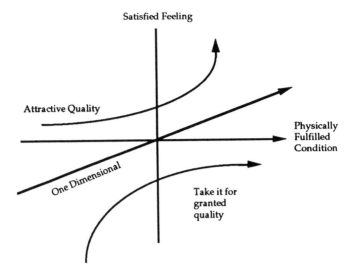

Figure 7.1. Customer satisfaction concepts.

or implied customer wants and needs. It requires the horizontal integration of those organizational functions that must Plan, Do, Check, and Act in order to successfully achieve customer-perceived expressed or unexpressed quality. Quality in this context is defined as the ability to meet or exceed customer expectations while maintaining a cost-competitive market position. Soichuro Honda was quoted to have said "As far as customers are concerned quality is something a product either has or does not have. There is no middle ground."

The key to success in QFD lies in ascertaining what the customer wants. Some call this obtaining the voice of the customer. This voice is fundamental to QFD as a planning process. The acceptance of this customer-driven element is what has enabled the users of QFD to attain a competitive quality, cost, or technical edge. One way to look at the impact of differential levels of customer importance can be illustrated by a model developed by Noriaki Kano* (see Figure 7.1).

The model defines three levels of customer satisfaction. Level one is termed *expected* or *taken for granted quality*. These are items that customers do not inform the supplier about but which, if they are omitted, will result in extreme customer dissatisfaction—for example, no soap in a hotel's rooms. This category is shown in the lower right quadrant of the Kano model (Figure 7.1)

---

* Dr. Noriaki Kano is a Japanese professor and consultant.

and indicates that these items can never be satisfiers but are significant dissatisfiers. The second level, that which extends in a straight line from the lower left to the upper right quadrant illustrates those items that the customer will describe, usually in some detail—an example would be a specification item. Here the customer is satisfied if the product performs and dissatisfied if it does not. The third level is known as *attractive quality* or *exciting quality* and represents opportunities for significant breakthroughs in customer satisfaction. Customers seldom express their wants or needs in this category except in vague terms. This may be due to a lack of knowledge regarding the item or because they have not thought about it. The items in this category are always satisfiers and never dissatisfiers.

It is very important in the development process that the project team should identify the expected quality items; these are quite often safety or comfort requirements that remain unspoken. If emphasis is placed solely on the specified items then the development process will go through many expensive iterations. It is imperative that adequate time and resources be applied to identifying customers and questioning them to obtain a complete, three-level, customer satisfaction opportunity list.

7.3 QFD AND DESIGN SYSTEMS

The old system of design spent a relatively short time on defining the product. The designers have little input from the customer and often the design is done in a vacuum. Different groups within research and engineering are given pieces of the project to design and do so at cross-purposes. The result of this process is miscommunication, missed schedules, delays, and disagreements. Key people get transferred, there is little documentation on the project, developers become nervous and make changes that result in more delays and redesigns. The old system of design is shown schematically in Figure 7.2. Too often in this system producing the product and redesign are started on the factory floor or at the customer's site. This constant process of design and redesign increases the cost of the product, causes customer complaints, and increases warranty costs.

A method of correcting this costly and demoralizing process is to have a clear understanding of the customer's needs and expectations as well as a

Figure 7.2. Old system of design.

THE FOUR PHASES OF QFD

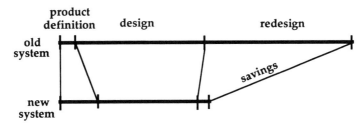

Figure 7.3. Comparison of old and new systems of design.

definitive product definition by involving all the needed organizations. This product definition should clearly spell out what the product is capable and not capable of doing in order to satisfy the voice of the customer.

The new system of design expands the time it takes to define the product but it shortens the design time by focusing priorities and better documentation, and improves communications, thereby significantly reducing the redesign and rework effort. The new system of design is compared to the old design system in Figure 7.3.

The old design system departmentalized the development cycle. This departmentalization resulted in barriers between the marketing, manufacturing, service, and design functions. These barriers caused inaccurate and conflicting communications. The communication problems amplified themselves into product delays that further increased the animosity between these functions. This in turn polarized the development process into finger-pointing, alienation, and alibiing. It is not uncommon in this type of system to spend more time fighting the system than developing the product.

The new design system utilizing QFD is based on obtaining the customer voice and defining customer demands and involves all required members of the organization. QFD makes design part of the organization's ongoing continuous improvement efforts.

7.4 THE FOUR PHASES OF QFD

QFD can be described as a four-phase approach to planning for design and development. The four phases are:

1. *The organization phase.* During this phase, management selects the product or service to be developed and the appropriate interdepartmental team. Management must decide on the extent of the QFD study. The extent of the study depends on the product or service complexity and the type of improvement sought. The people involved with the study

need to identify who the study is for; how much funding is being provided; what the expected time frame is for the project. When selecting a project, these questions must be answered: Why the project is being done, what breakthroughs are needed, and where the breakthroughs are likely to be?

2. *The descriptive phase.* The selected project team describes the product or service in terms of actual customer words, customer demands, functions, cost, reliability, and characteristics.

The descriptive phase begins with an investigation of the voice of the customer. It is important to understand "what" are the customer wants and needs and establish measures to ensure fulfillment. During this phase it must be decided who is meant by the term *customer*. Is the customer external to the company or internal? Is the customer an end-user, a retailer or a wholesaler; a distributor, a subsidiary, a contractor? This is important because different customers have differing wants and needs that are peculiar to their role in a business sector in addition to differences between individuals or groups within the same sector.

In the descriptive phase one is gathering and analyzing information to provide pointers as to the most significant customer demands, and their associated producer-controllable critical quality characteristics, that will guide one to breakthrough opportunities and design improvements.

Customer demands should be understandable within the producer organization in order to allow identification of the controllable and measurable items needed to meet the requirements. Customer demands should:

Be traceable to customer words
Be precise, for example: verb plus a modifier—"easy to use"; noun plus adjective—"lasts a long time" or "small package"
Avoid using numbers
Use positive statements

The producer-controllable items that will enable meeting of customer demands are commonly called *quality characteristics* or *counterpart characteristics* and in some cases *design requirements*. The key to these characteristics is that they should be measurable—for example, temperature, weight, height, responsiveness, time, density, or readability. There should be at least one characteristic for each customer demand. However, each characteristic may have a relationship to other than its prime customer demand.

THE FOUR PHASES OF QFD

3. *The breakthrough phase.* In this phase the team selects areas for improvement and competitive advantage through the investigation of new technology, new concepts, improved reliability, cost reduction, and bottleneck investigation.

This investigation is accomplished by using an appropriate matrix or matrices from the matrix of matrices chart, as shown in Figure 7.4. The appropriate matrix is chosen on the basis of the purpose to be achieved. The matrix relationships are shown in Figure 7.5. The A-1 matrix will be explained in more detail in the illustrative example.

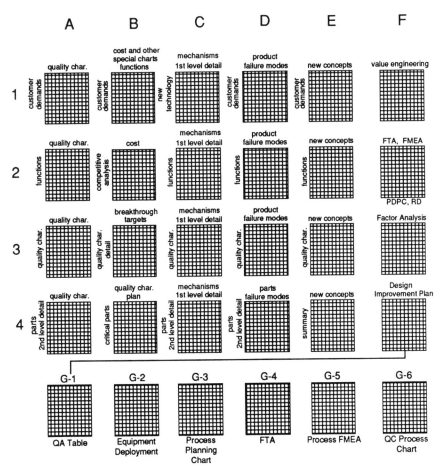

Figure 7.4. Matrix of matrices (reprinted with permission from GOAL/QPC).

154 CUSTOMER-DRIVEN ENGINEERING. QUALITY FUNCTION DEPLOYMENT

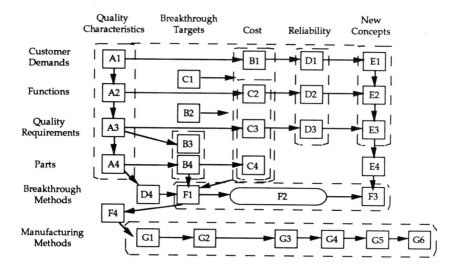

Purpose to be achieved	Charts to Use
Analyze customer demands	A1, B1, D1, E1
Critique functions	A2, C2, D2, E2
Set quality characteristics	A1, A2, A3, A4
	B3, B4, C3, D3, E3
Identify critical parts	A4, B4, C4, E4
Set breakthrough targets	C1, B2, B3, B4
Set cost targets	B1, C2, C3, C4
Set reliability targets	D1, D2, D3, D4
Select new concepts	E1, E2, E3, E4
Identify breakthrough methods	D4, F1, F2, F3
Identify manufacturing methods	G1, G2, G3, G4, G5, G6

Figure 7.5. Matrix relationships.

4. *The implementation phase.* The project team describes and defines the product or service and establishes how it will be produced and how the process will be controlled.

The following steps are a general outline to select and facilitate a QFD project successfully:

Selections
- Identify projects that support company priorities.
- Select projects that will improve key interfaces.

THE FOUR PHASES OF QFD

- Involve all essential departments needed to accomplish the objectives.
- Involve personnel who believe QFD will work.
- Select projects that are likely to succeed.
- Select projects that are likely to have an economic impact.

Facilitators
- Clearly define the project.
- Obtain management commitment to taking action on the findings.
- Focus on the process of QFD rather than the content of the project.

Select the team
- A rule of thumb is to have from three to eight people who possess the required and complementary expertise.

Select the leader
- The leader should be an individual who has a relevant role in the design and delivery of the project.

Provide a mandate
- The team must be provided with the three "Ps"—permission, power, and protection.

Provide training
- Make sure that each project team member has the basic skills and knowledge to function in the planning process.

Provide facilitator
- The facilitator needs an in-depth understanding of QFD and the seven management and planning tools.

The Voice of the Customer

There are many ways to obtain the voice of the customer. Some of the more likely methods include market surveys, focus groups, and interviews. In any of the techniques it is important to remember that the aim is to listen to the customer to obtain "customer verbatims." This is the basic starting point and is vital to the success of a QFD effort. Brainstorming via a method known as the affinity diagram and use of a tree diagram, two of the seven management and planning tools, are two of the more successful methods of expanding and translating customer words into usable customer demands. Continually check back to the original customer words to make sure the meaning has not changed during the translation process.

7.5 QUALITY FUNCTION DEPLOYMENT CASE STUDY

The Double M Electronics Corporation is currently preparing to start on a QFD study on its wave solder process. The management of the corporation has selected this as the first project to utilize the QFD process and is anxious to have a success. Management is now putting together the appropriate interdepartmental team members and appointing a program manager to lead the project and the QFD study and a facilitator versed in QFD.

The team will consist of decision-making-level representatives of manufacturing, design, quality, field service, and the program manager as the team leader. This team is structured by management to have the appropriate cross-functional representation and authority to successfully complete the QFD study in the solder process. The team has decided to enlist the services of an experienced QFD facilitator to lead them through the process.

The QFD process begins with the identification of the customer and what the customer desires from the soldering process—the voice of the customer. Who is the customer—Internal or external?

The first step for the team is to develop a listing of customer demands. Customer demands are statements of needs and wants from internal and external customers. The team accomplished this step by brainstorming a list of customer demands for the solder process. The team members each had, in the past, numerous contacts with their customers and had a sufficient knowledge base to create an accurate list of demands. After the demands were listed, they were organized into a matrix diagram as shown in Figure 7.6. The facilitator next explained the usage of the A1 matrix defining the chart's input, output, and benefits as shown in Figure 7.7.

The facilitator then had the team rank the importance of these customer demands based on their experience and interaction with the current customer base. The rating scale is:

1 Very low importance to the customer
2 Low importance to the customer
3 Some importance to the customer
4 Important
5 Very important to the customer

The team developed the customer importance rating for the lowest level of detail on the customer demand matrix through consensus, as shown in Figure 7.8.

The next step was to determine where the company was in relation to the customer importance ranking through the team consensus process, as shown in Figure 7.9.

QUALITY FUNCTION DEPLOYMENT CASE STUDY   157

The team must now complete the competitive analysis section to determine how well their competition is doing in satisfying the customer base they are evaluating. This process step is accomplished by a combination of experience, field visits, product comparisons, and/or surveys. The benchmarking scores are developed through consensus as shown in Figure 7.10, using the same scale to rate the importance of the customer demands.

Now the team must develop the company's improvement plan. This step must carefully consider the following inputs:

Importance to customers
Competition
Marketing window

The team, through a consensus process, arrived at the improvement plan figures shown in Figure 7.11.

The calculation of the ratio of improvement column shown in Figure 7.12 is calculated by dividing the plan by the "company now" for each row. For example: for "free of opens":

$$\text{Ratio of improvement} = \frac{\text{Company plan}}{\text{Company now}} = \frac{5}{3} \approx 1.7$$

The next step for the team, as shown in Figure 7.13, was to determine whether any sales points existed. A sales point indicates a quality characteristic that would excite our customer(s) and would likely be used in a advertising program. Sales points can be characterized as follows, with the ratings indicated:

Super sales point	Customer must have it and will pay to have it	1.5
Interesting sales point	Customer would be interested but concerned with cost	1.2
Ho-hum sales point	Nothing new	1.0

The last two columns to be computed on the right-hand side of the A-1 chart are the absolute weight and demanded weight. The *absolute weight* is computed by multiplying the rating of importance by the ratio of improvement by the sales point. The *demanded weight* is calculated by totaling the absolute weight and converting the individual absolute weights to percentages of the total. This calculation is shown in Figure 7.14. For example, for "free of opens,"

Rate of importance × Rate of improvement × Sales point = Absolute weight

     5   ×   1.7   ×   1.5   =   12.75

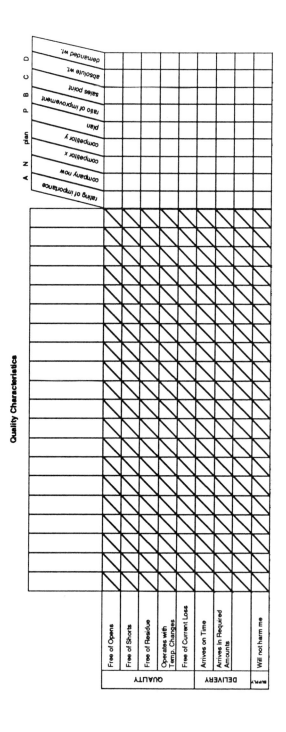

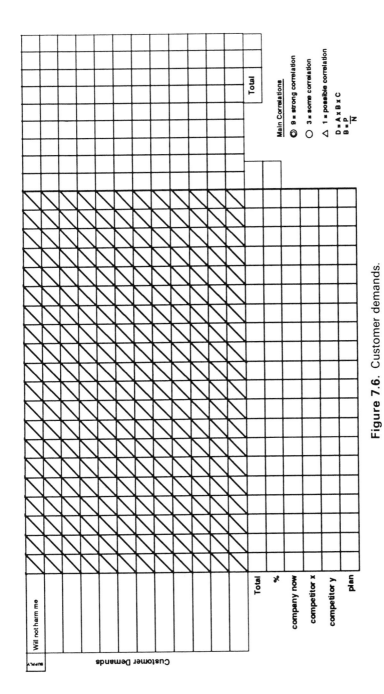

Figure 7.6. Customer demands.

Chart	Inputs	Outputs	Benefits
A-1 Quality Characteristics vs. Customer Demands "Voice of the Customer"	• Customer Demands • Quality Characteristics • Company Now • Competitor's Results	• Which demands are most important to customer • How company compares to the competition • Most significant sales points • What are the key Quality characteristics necessary to meet key cost and demands	• Identifies potential breakthrough areas with greatest appeal to customer

Figure 7.7. Chart A-1.

This part of the A-1 chart indicates to the QFD project team the top four customer demands that are the most important on the basis of importance to customers, the company improvement ratio, and the sales point calculations.

The facilitator explained to the team that the top horizontal part of the chart was the next step in the process: the team would now develop the quality characteristics. Quality characteristics are quality elements that are measurable and controllable characteristics by which we meet customer demands. During this phase we avoid specifying test parameters but instead use the test purpose. The team brainstormed a listing of quality characteristics and arranged them as shown in Figure 7.15.

The quality characteristics should answer the question of what measurements will ensure that the customer demands are met. In this step, avoid specifying test parameters, means of accomplishment, cost, or reliability items. Quality characteristics should be items that are measurable and controllable. The team should ask of each quality characteristic: "If I measure/control———, I will meet this customer demand"

Once the team had listed the quality characteristics on the chart, they then began to correlate them to the customer demands as shown in Figure 7.16. The team used the following symbols to represent the level of correlation between a customer demand and a quality characteristic:

⊙ Strong correlation = 9
○ Some correlation = 3
△ Possible correlation = 1

The numerical value associated with the symbol is an indication of the strength of the relationship.

Now the team was ready to calculate the total for each quality characteristic. This calculation is accomplished by multiplying the correlation values by the demanded weight as shown in figures. The columns are then totaled and

QUALITY FUNCTION DEPLOYMENT CASE STUDY

each column is converted to a percentage of the total as shown in Figure 7.17. For example,

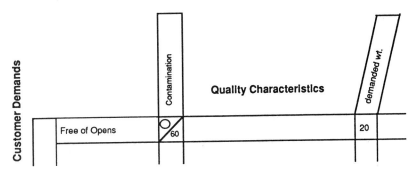

(Remember that an open circle has a value of a 3.) The calculation is the correlation weight (3) × demanded weight (20) = 60.

The team was then able to rank order the quality characteristics and identify the key item for the design engineers to work on. The rank ordering of the quality characteristics for the solder process was:

1. Contamination
2. Yield
3. {Timelines
   {Temperature
4. Travel time

The last step is to list the target values for each measurable quality characteristic for the company and its competitors, as shown in Figure 7.18.

The completed A-1 chart provides a picture of how the customer demands and quality characteristics interact with each other. It also gives a competitive picture and points out current strengths and weaknesses.

Since it is beyond the scope of this book to go beyond chart A-1, a recommended sequence for the team to follow to complete the design phase of this project is as follows:

A-1   Customer demands/quality characteristics
B-1   Customer vs. functions
A-2   Functions/characteristics
C-2   Functions/conceptual mechanisms
C-4   Real mechanisms/parts
D-4   Parts/parts failure modes

This is a recommended sequence for an organization to take for the first project in QFD.

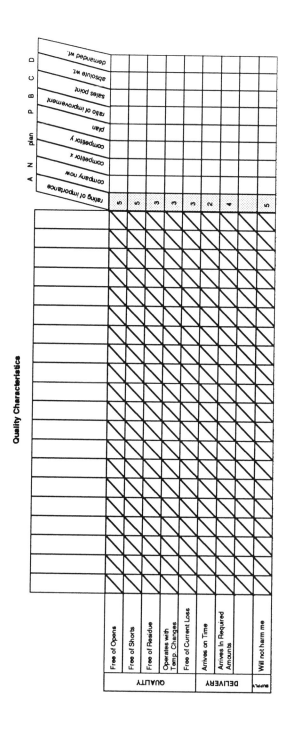

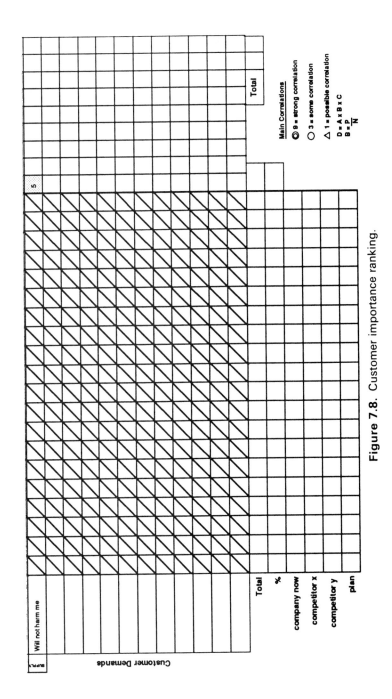

Figure 7.8. Customer importance ranking.

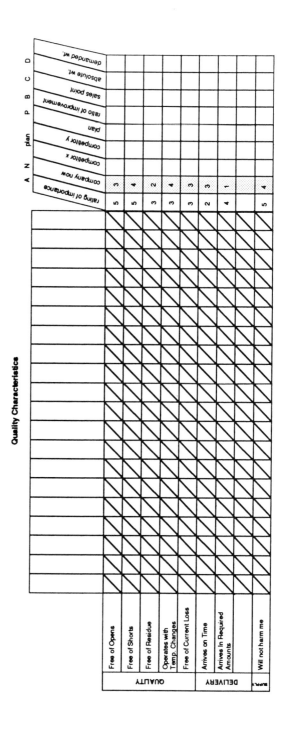

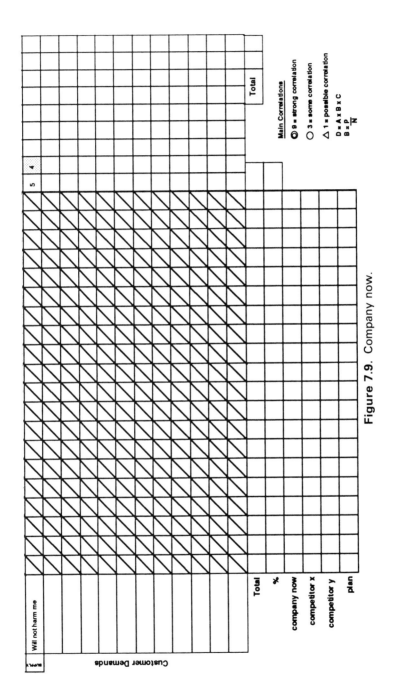

Figure 7.9. Company now.

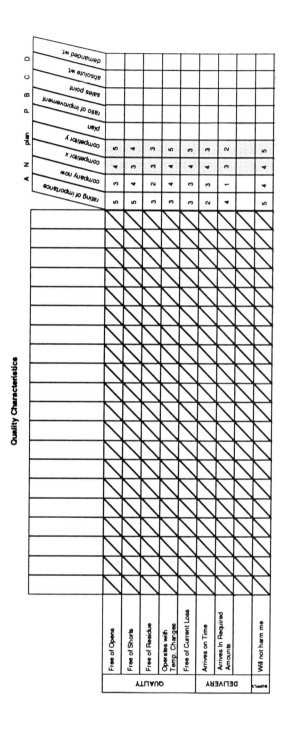

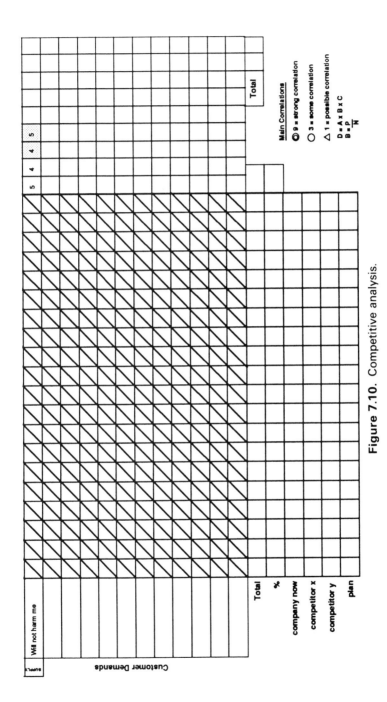

Figure 7.10. Competitive analysis.

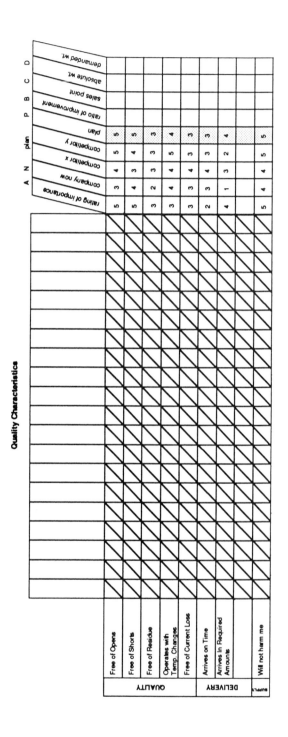

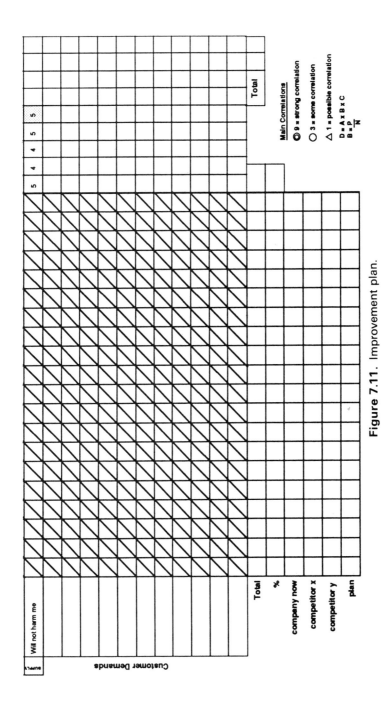

Figure 7.11. Improvement plan.

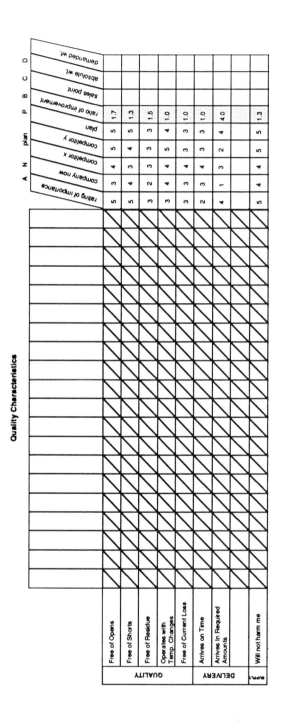

Quality Characteristics

		rating of importance	A company now	N competitor x	competitor y	plan plan	P ratio of improvement	B sales point	C absolute wt.	D demanded wt.
QUALITY	Free of Opens	5	3	4	5	5	1.7			
	Free of Shorts	5	4	3	4	5	1.3			
	Free of Residue	3	2	3	3	3	1.5			
	Operates with Temp. Changes	3	4	4	5	4	1.0			
	Free of Current Loss	3	3	4	3	3	1.0			
DELIVERY	Arrives on Time	2	3	4	3	3	1.0			
	Arrives in Required Amounts	4	1	3	2	4	4.0			
SUPPLY	Will not harm me	5	4	4	5	5	1.3			

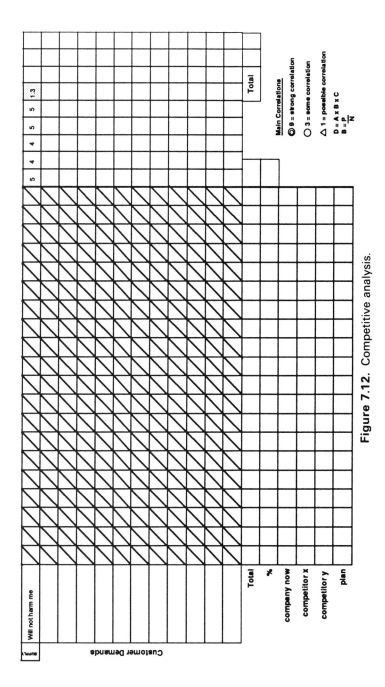

Figure 7.12. Competitive analysis.

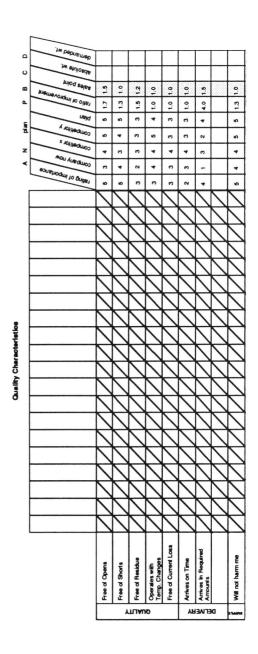

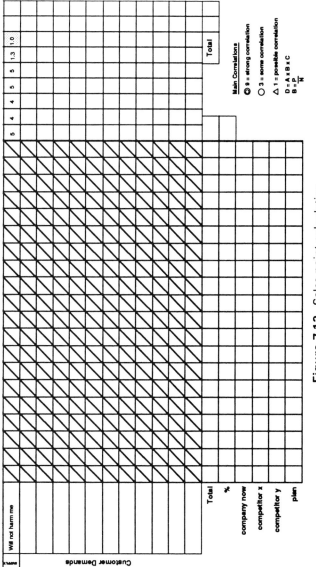

Figure 7.13. Sales point calculation.

Quality Characteristics

		rating of importance	company now	competitor x	competitor y	plan	ratio of improvement	sales point	absolute wt.	demanded wt.
			A	N		plan	P	B	C	D
QUALITY	Free of Opens	5	3	4	5	5	1.7	1.5	12.75	20
	Free of Shorts	5	4	3	4	5	1.3	1.0	6.5	10
	Free of Residue	3	2	3	3	3	1.5	1.2	5.4	9
	Operates with Temp. Changes	3	4	4	5	4	1.0	1.0	3.0	5
	Free of Current Loss	3	3	4	3	3	1.0	1.0	3.0	5
DELIVERY	Arrives on Time	2	3	4	3	3	1.0	1.0	2.0	3
	Arrives in Required Amounts	4	1	3	2	4	4.0	1.5	2.4	38
SUPPLY	Will not harm me	5	4	4	5	5	1.3	1.0	6.5	10

174

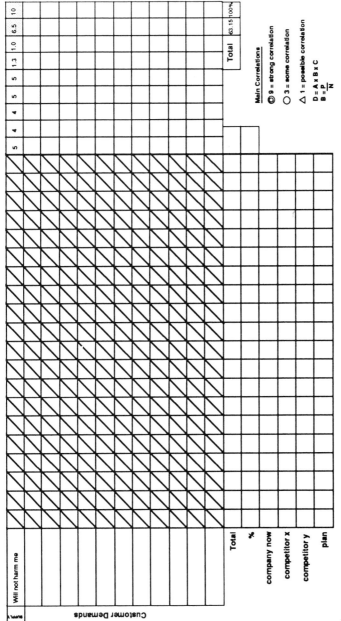

Figure 7.14. Absolute and demanded weight.

Quality Characteristics

Quality characteristics (columns): Contamination, Temperature, Yields, Timeliness, Workability, Travel Time

		rating of importance	A company now	N competitor x	competitor y	plan	P ratio of improvement	B sales point	C absolute wt.	D demanded wt.
QUALITY	Free of Opens	5	3	4	5	5	1.7	1.5	2.75	20
	Free of Shorts	5	4	3	4	5	1.3	1.0	6.5	10
	Free of Residue	3	2	3	3	3	1.5	1.2	5.4	9
	Operates with Temp. Changes	3	4	4	5	4	1.0	1.0	3.0	5
	Free of Current Loss	3	3	4	3	3	1.0	1.0	3.0	5
DELIVERY	Arrives on Time	2	3	4	3	3	1.0	1.0	2.0	3
	Arrives in Required Amounts	4	1	3	2	4	4.0	1.5	2.4	38
SUPPLY	Will not harm me	5	4	4	5	5	1.3	1.0	6.5	10

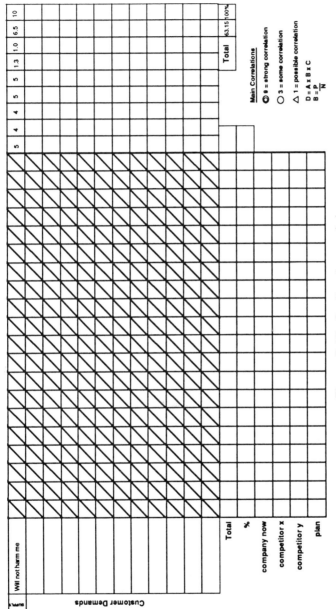

Figure 7.15. Quality characteristics.

Quality Characteristics

		Contamination	Temperature	Yields	Timeliness	Wickability	Travel Time	rating of importance	company now	competitor x	competitor y	plan	ratio of improvement	sales point	absolute wt.	demanded wt.
									A	N		P		B	C	D
QUALITY	Free of Opens	◉		◉	○		○	5	3	4	5	5	1.7	1.5	12.75	20
	Free of Shorts	○		○	○		△	5	4	3	4	5	1.3	1.0	6.5	10
	Free of Residue	◉	◉					3	2	3	3	3	1.5	1.0	5.4	9
	Operates with Temp. Changes	◉				△		3	4	4	5	4	1.0	1.2	3.0	5
	Free of Current Loss	○						3	3	4	3	3	1.0	1.0	3.0	5
DELIVERY	Arrives on Time	○	◉	○	◉			2	3	4	3	3	1.0	1.0	2.0	3
	Arrives in Required Amounts	○		◉	○			4	1	3	2	4	4.0	1.5	2.4	38
SUPPLY	Will not harm me	△	○				○	5	4	4	5	5	1.3	1.0	6.5	10

178

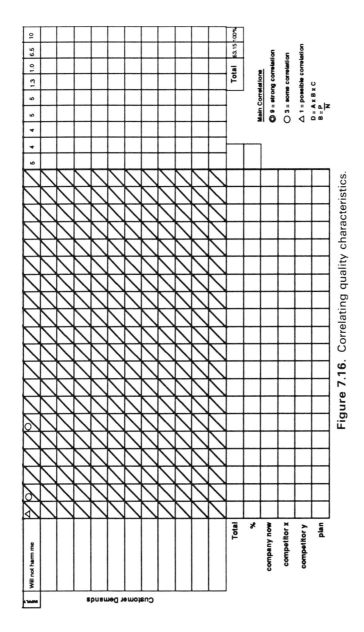

Figure 7.16. Correlating quality characteristics.

Quality Characteristics

		Contamination	Temperature	Yields	Timeliness	Wickability	Travel Time			rating of importance	company now	competitor x	competitor y	plan	ratio of improvement	sales point	absolute wt.	demanded wt.
											A	N			P	B	C	D
QUALITY	Free of Opens	◎ 180	◎ 30	◎ 60	◎ 60		○ 60			5	3	4	5	5	1.7	1.5	12.75	20
	Free of Shorts	○ 30	◎ 90	○ 30	○ 30		△ 10			5	4	3	4	5	1.3	1.0	6.5	10
	Free of Residue	◎ 81				△ 9				3	2	3	3	3	1.5	1.2	5.4	9
	Operates with Temp. Changes		◎ 45							3	4	4	5	4	1.0	1.0	3.0	5
	Free of Current Loss	◎ 45		○ 9						3	3	4	3	3	1.0	1.0	3.0	5
DELIVERY	Arrives on Time	○ 9		◎ 9	◎ 27					2	3	4	3	3	1.0	1.0	2.0	3
	Arrives in Required Amounts	○ 114		◎ 342	○ 114					4	1	3	2	4	4.0	1.5	2.4	38
SUPPLY	Will not harm me	△ 10	○ 30				○			5	4	4	5	5	1.3	1.0	6.5	10

180

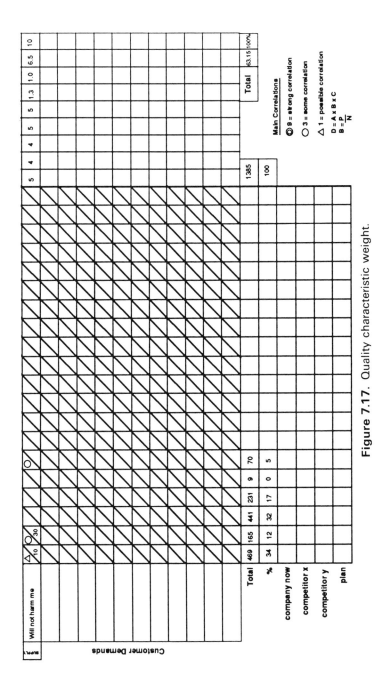

Figure 7.17. Quality characteristic weight.

		Contamination	Temperature	Yields	Timeliness	Workability	Travel Time				rating of importance	company now	competitor x	competitor y	plan	ratio of improvement	sales point	absolute wt.	demanded wt.
												A	N		plan	P	B	C	D
QUALITY	Free of Opens	◎ 180	◎ 90	◎ 60	◎ 60		○ 60				5	3	4	5	5	1.7	1.5	12.75	20
	Free of Shorts	30	30	○ 30	○ 30		△ 10				5	4	3	4	5	1.3	1.0	6.5	10
	Free of Residue	◎ 81				△ 9					3	3	3	3	3	1.5	1.2	5.4	9
	Operates with Temp. Changes		◎ 45								3	4	4	5	4	1.0	1.0	3.0	5
	Free of Current Loss	◎ 45									3	3	3	3	3	1.0	1.0	3.0	5
DELIVERY	Arrives on Time	○ 9		○ 9	◎ 27						2	3	3	3	3	1.0	1.0	2.0	3
	Arrives in Required Amounts	114		◎ 42	○ 114						4	1	3	2	4	4.0	1.5	2.4	38
SUPPLY	Will not harm me	△ 10	○ 30				○				5	4	4	5	5	1.3	1.0	6.5	10

Quality Characteristics

182

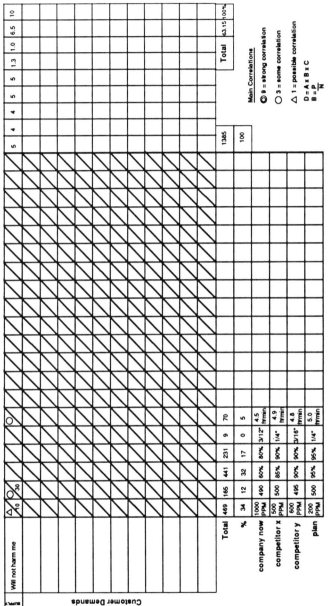

Figure 7.18. Quality characteristic target value.

7.6 CONCLUSION

A successful QFD implementation into an organization's design process of new product development will allow the organization to:

Build a data base for future designs
Provide a project tracking system
Enhance communication
Foster team building
Reduce quality cost
Increase customer satisfaction

In order to have a successful application of QFD within an organization, top management must:

Make it clear that QFD is a priority
Set clear priorities for QFD activities
Make training available
Insist that designs be based upon customer requirements
Become leaders of QFD rather than managers

The authors have learned from their teaching and consulting experience that successful organizations using QFD initially followed these guidelines:

The first projects should be on existing products/services.
QFD works best focused to help you solve a problem.
Involve people that are interested in QFD for the initial projects.
Do not let the A-1 chart get too large.
Try to use a variety of charts in the initial projects.

Those organizations not having a successful implementation of QFD usually made the following mistakes:

Charts that are too big
Use of a A-1 only
Mixing engineering demands (functions) with customer demands on A-1
QFD completed too late to implement changes
No buy-in for QFD-suggested changes
Identifying means as quality characteristics

In addition to these mistakes, many organizations have doubters and detractors who usually raise the following opposition points to QFD:

Payoff is not clear.
QFD needs long term thinkers.
People don't understand it.

People are not looking for another task.
It is just another program.
It is just a fad.
Resources are limited.
No more training is possible.
The company does not have the time.

QFD is a disciplined process that if implemented correctly in an organization will provide the following benefits:

Listening to the voice of the customer
Improving horizontal communications
Prioritizing improvements
Targeting cost reduction
Targeting reliability projects
Targeting engineering breakthroughs
Orchestrating engineering breakthroughs
Improving communication between design and manufacturing

QFD is not a quick fix to an ailing design system. It is a process that requires a long-term approach, since it requires an organizational transformation from the product-out design mentality to the market-in design approach. The voice of the engineer is replaced with the voice of the customer in the design through manufacturing process.

7.7 GLOSSARY OF QFD TERMS

Breakthrough targets Identified process or product characteristics that will provide a competitive or market edge if implemented.

Conceptual mechanisms The major subgroups sometimes referred to as *first level of detail* upon which the QFD study will focus.

Customer words Untranslated written or oral expressions received from surveys, audits, and other sources that describe customer wants.

Customer demands A producer translation of customer words that can be understood by most people in the development cycle.

Design of experiments A mathematical technique that enables the identification of the most probable causes of problems in a multivariable environment.

Flow chart A pictorial representation that identifies each step involved in developing and producing a product or service.

FMEA An abbreviation for *failure mode and effect analysis* that indicates the priority of failures to work on and indicates severity, probability, and criticality of occurrence.

FTA An abbreviation for *fault tree analysis*—a treeing technique that shows the interrelationship between causes of failure.

Function A description of what a product or service does. It is normally expressed in the form of subject–verb–object.

New concepts Different or innovative ways of thinking about the product or service.

Parts A subgroup of mechanisms or first level of detail. The second level of detail is those items that enable a mechanism to become operable.

Parts failure modes Ways in which the items that make up the product or service mechanisms can fail.

Potential failure cause A prior assessment of the most likely contributors to a failure.

Potential failure effect A risk assessment of the impact of a future failure on a product or service.

Potential failure mode A preproduction problem analysis activity.

Process failure modes Ways in which the process that produces the product or service can fail.

Product failure modes Ways in which the product or service can fail.

QA table A summary that enhances communication of important product items between design and manufacturing.

QC process plan chart Identifies by name, number, and description how the process will be controlled.

QFD An analytical system that improves horizontal communication. A planning method to develop products or services based upon identifying and satisfying customer demands.

Quality characteristics Those measurable/controllable parameters or techniques that when properly implemented will ensure that the customer demands are met.

Sales point An identified customer demand that if satisfied will enable a significant improvement in sales or market position.

SUGGESTED READING

Akao, Yoji (ed.). *Quality Function Deployment: Integrating Customer Requirements Into Product Design.* Cambridge, Mass.: Productivity Press, 1990.

SUGGESTED READING

Akao, Y., Ohfuji, T. and Naoi, T. "Surveys and reviews on quality function deployment in Japan." *ICQC Proceedings*. Methuen, Mass.: GOAL/QPC, 1987.

Cohen, Louis. "Quality function deployment: An application perspective from Digital Equipment Corporation." *National Productivity Review*, pp. 197–208, Summer 1988.

Daetz, Doug, Flaherty, Thomas, Kotecki, Mary Lou, Mazur, Glen, Marsh, Stan, Moran, John and ReVelle, Jack. *Quality Function Deployment: A Process of Continuous Improvement*. Research Report #89-10-02. Methuen, Mass.: GOAL/QPC, 1989.

DeVera, Dennis, Glennon, Tom, Kenny, Andrew A., Khan, Mohammad A.H. and Mayer, Mike. "An automotive case study." *Quality Progress*, Vol. XXI, No. 6, pp. 35–38, June 1988.

Eureka, William and Ryan, Nancy. *The Customer Driven Company*. Dearborn, Mich.: ASI Press, 1988.

Fortuna, Ronald M. "Beyond quality: Taking SPC upstream," *Quality Progress*, Vol. XXI, No. 6, pp. 23–28, June 1988.

Hauser, John R. and Clausing, Don. "The house of quality." *Harvard Business Review*, May–June, No. 3, pp. 63–73, 1988.

Kenny, Andrew A. "A new paradigm for quality assurance." *Quality Progress*, Vol. XXI, No. 6, pp. 30–32, June 1988.

King, Robert. "Listening to the voice of the customer: Using the quality function deployment system." *National Productivity Review*, pp. 277–281, Summer 1987.

King, Robert. *Better Designs in Half the Time: Implementing QFD in America*. Methuen, Mass.: GOAL/QPC, 1987.

Kogure, Masao and Akao, Yoji. "Quality function deployment and company-wide quality control in Japan." *Quality Progress*, Vol. XVI, No. 10, pp. 25–29, Oct. 1983.

Moran, John W. "Lessons learned in applying QFD." Transactions of the 1989 Symposium on QFD, Automotive Division, ASQC.

Ross, Phillip J. "The role of Taguchi methods and design of experiments in QFD." *Quality Progress*, Vol. XXI, No. 6, pp. 41–47, June 1988.

Sullivan, L.P. "Quality function deployment." *Quality Progress*, Vol. XIX, No. 6, pp. 39–50, June 1986.

Sullivan, Lawrence P. "Policy management through quality function deployment." *Quality Progress*, Vol. XXI, No. 6, pp. 18–20, June 1988.

CHAPTER 8

The Manufacturing Process and Design Ratings

Design ratings are becoming important in formalizing the design for manufacture (DFM) guidelines and choosing between competitive or alternate designs. This chapter will discuss the current methodologies of design guidelines, and emphasize their differences and the correct application of each.

Design ratings can serve as the common language of both design and manufacturing engineers, and have been especially embraced by manufacturing as the primary vehicle in which to express their concerns for improving DFM. Training in these methods involves both engineering communities and has created an atmosphere that fosters DFM in the total organization. Design ratings could also be used as a training method for new engineers to become more attuned to manufacturing processes and techniques.

Most of the design ratings methods originated as guidelines for manual assembly, and have since matured to include automatic and flexible assembly, sheet metal, plastic molding, and printed circuit boards. They have been quickly adopted by major manufacturing companies with a high degree of automation such as the ones in the automobile and personal-computer industries. For this reason, there is a general feeling that these design ratings methods are only beneficial to large-volume manufacturing. However, they could be even more important to small-volume industries such as military, space, and medical device manufacturers because they emphasize reducing the costs of manufacturing and field support of products, since individual unit costs are very high. In addition, the cost of special tools and spare parts has also increased significantly.

8.1 THE MANUFACTURING PROCESS FOR ELECTRONIC PRODUCTS

The majority of electronic products made in the United States are not consumer products but are low-volume devices made for defense, medical, and space applications, and the support equipment for the design, manufacture, and test of electronic components and products. Production volumes vary for different industries, from hundreds per month for medical, space, and defense industries, to thousands per month for machine-tool control, engineering workstations, and test equipment, to tens of thousands per month for personal computers and their peripherals. Consumer, entertainment, and automobile industry electronic assemblies are also manufactured at very high volumes.

Though recent technological improvements have increased the integration of electronic components, the design of electronic products remains basically the same. Information is being input through sensors, keyboards, or from other electronic systems, processed electronically, and then displayed to the user through printed or graphical means or transmitted to other electronic products. The majority of the manufacturing labor is consumed in the final assembly of the electronic components and subsystems, and this is the area targeted for the greatest benefit from design for manufacturing.

8.1.1 Printed Circuit Boards

Printed circuit boards (PCBs) are used as the vehicle for connecting the individual electronic components. The PCBs are then connected together to form a central system for power and electronic signal distribution, through a backplane standard connection, referred to as the *motherboard*. There have been many techniques in the ingenious use of PCBs through extended surface features that act as the input or output signal receivers for the product. The PCBs are fabricated and assembled in standard processes, with specific automation equipment available to enhance their quality and reduce their costs.

The concurrent engineering and design for manufacture efforts for PCB design have been focused on creating a standard PCB process, both for fabrication and assembly, in which most of the company designs can be implemented. These standard processes have been documented through several societies and engineering concerns, especially the Institute of Printed Circuits (IPC) of Lincolnwood, Illinois. It is well understood that if the design engineers follow the guidelines for these standard PCB processes, the quality and cost of the PCBs in their products will be optimized.

One of the difficult problems facing design engineers is the need for developing special PCB processes that are then adopted as part of the company's competitive position. These needs could arise from environmental or user-related situations and might include high resistance and low leakage, conformal coatings, temperature and dielectric properties, and shock and vibrations resistance. Unless absolutely necessary, these special conditions should be avoided, because they lock the manufacturing operation into specialized and difficult-to-manufacture processes. In addition, the manufacturing costs of these special PCBs and their quality cannot be compared to similar standard operations, so that their true competitive advantage cannot be fully evaluated. Special PCB manufacturing operations tend to restrict new designs and inhibit engineers from exploring new methods for interconnections.

PCBs have created their own industry for automating both the fabrication and assembly processes. The suppliers of the PCBs tools and equipment are very similar in their basic equipment designs and technologies. They advertise their wares through technical magazines and exhibits, the largest being the National Electronic Packaging and Production Conference (NEPCON). The equipment suppliers compete on features, and will leapfrog each other in terms of new capabilities through the evolving generations of equipment.

The PCB equipment suppliers have created good DFM guidelines for their customers in order to maximize the use of their equipment. However, with the advent of computers and microprocessors, many equipment controls have become overly flexible in terms of their settings, with more parameters open to adjustments by users. This increased flexibility in equipment adjustments has led to the opposite of the intended effect, since users are confused about customizing their processes to fit their PCBs. Design for manufacture of PCBs will be discussed in detail in Chapter 10.

8.1.2 Sheet Metal Fabrication

Sheet metal is the preferred packaging method for low-volume production of electronic products because of its low tooling costs. Tight tolerances can now be achieved with sheet metal through laser cutting and other technological innovations. Such advances in sheet metal technologies as flexible manufacturing systems (FMS) for punching, shearing, and bending, automatic material conveyance through automatic ground vehicles (AGVs), and robotic welding and powder painting are further reducing the setup and therefore the cost of sheet metal operations. Another inherent advantage of sheet metal is that it can act as a conductive shield for grounding and inhibiting interference radiation of electronic products.

The issues of designing for sheet metal operations are based on the tolerances of the finished piece, the geometry of the sheet metal design and the assembly methods required. The tolerances are the summation of those of the raw materials and the fabrication processes. Repeatability of the sheet metal parts produced will be dependent on the designer's awareness of these tolerances. It is desirable to have the fabricator make all complex shapes, rather than to build them up from multiple pieces.

The geometry of the sheet metal design has inherent costs associated with it, the relationship of the single part to sheet size is important in determining the number of parts that can be made from a single sheet. In addition, the number, sizes, and shapes of the holes to be punched out determines the time allotted for the punching operation, and therefore the cost of the part. The size of the sheet metal piece also determines the need for stiffeners for strength and rigidity.

The assembly method is important for the design's efficiency. Typical assembly methods are spot, tack, or continuous welding, depending on the functionality or finish required. Spot welding is preferred because it is fast, and produces the least amount of distortion and surface blemish. Tack welding is done inside the part and on rear corners where appearance is not important. Continuous welds are the most expensive to apply and remove, and cause the most distortion in metal parts.

In order to reduce tooling costs, it is desirable to make the parts interlocking and self-aligning by using notching and indexing mating parts or by providing a locator on which an adjoining part can be positioned without the assistance of clamps or stops.

8.1.3 Plastic Parts

Plastic parts can provide the best packaging, by being used as the chassis to hold the products, with close tolerances and minimum unit costs. Several complex contours and shapes can be achieved that are not possible with sheet metal. Tooling costs for plastic parts are expensive, and a large volume is required to justify them.

There is a long development cycle and different construction for plastic tooling, ranging from permanent castings with long production runs to investment castings, which are transfer methods for making parts for limited production runs. Recently, new technologies of laser thermal curing called *stereo lithography* have increased the prospects of making prototype plastic parts instantly for trying out the design geometry before committing to hard tooling. The design can be made from CAD drawings directly by using thermally cured plastic compounds, and the parts that are produced are made one layer at a time, complete with internal holes and recesses, and with close enough tolerances to allow for testing of snap fits and interlocks. The equipment available is expensive, but prototypes can be made from local suppliers of the technology.

The advantage of plastic parts is that complex geometry can be achieved with a minimum effect on the production cost, and the geometry influences only the one-time charge of tooling. The design guidelines for increasing efficiency can be easily achieved with plastic parts by making them symmetrical and easily assembled using snap fits and thread-forming of screws and inserts.

8.1.4 The Assembly Process

The assembly process can be achieved by one of three different methods: manual, hard automation, and flexible assembly or robotics. Figure 8.1 demonstrates the typical application areas for the three assembly types versus annual production volume. The unit cost per assembly is also dependent on the annual production volume and the type of assembly used. It will remain constant for manual assembly, decrease linearly with increasing volume for automatic assembly, and behave as a hyperbolic function for robotic assembly.

The manual assembly process is adaptable, because people are very flexible and ingenious at making improvements in manual assembly work. The training of assembly workers is a very important factor in producing high-quality work. Another factor is the provision of good leadership through

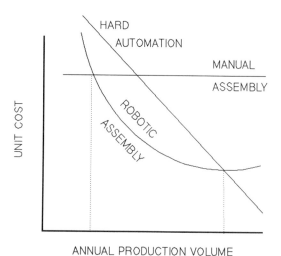

Figure 8.1. Comparison of assembly systems.

supervisory and management personnel. The start-up cost of a manual assembly operation is very low, and expansion is simple and inexpensive, requiring only additional factory space to put together more benches and workstations. The assembly cost of manual assembly is constant versus production volumes, and manual assembly productivity tends to keep up marginally with wage inflation, because the rate of production is difficult to increase since speed and accuracy are limited by human physical skills.

Hard automation was the main system of assembly automation before the advent of robotics and flexible automation. It involves special-purpose machines that are used to locate, orient, and grasp the parts, transfer them to an assembly station, and then perform the joining operation. These systems require specialized machine designers and skilled technicians to maintain and support the equipment. They are special-purpose in that they perform unique operations in assembling particular parts. In the event of an assembly upgrade or redesign, extensive modifications may have to be performed on the equipment. They feature fixed cycle time and constant, high production rate. The part unit costs will decrease as the annual volume increases because the automation investment is spread over a larger number of parts. When automating an assembly process, it is important that the specifications, quality, and manufacturing processes of the parts are well documented, as the automatic production rate could be seriously curtailed if poor-quality or defective parts were present in the process.

Flexible or robotic assembly is a good compromise between the speed and accuracy of hard automation and the flexibility of manual assembly. The equipment is usually adaptable to design changes, and can be flexible and repeatable in production volumes. The technical support required to maintain the equipment and the process is fairly routine, with robots being programmed either by using a movement-path memorizer pendant or by writing a program in a BASIC-like language to perform the operations.

Robots can be purchased in integrated systems, with parts presented using specially designed trays programmed manually or automatically. There is a multiplicity of robotic arms and workheads, with different degrees of freedom and axes of movement. The most common types are pick-and-place operations that can grasp parts from a predetermined position, carry them over to the assembly point, position the parts for assembly, and then perform the joining operation.

Although robotic assembly has yet to make a major impact on medium-volume electronic assembly, it has been used in very specific applications in the electronic industry: hazardous and dangerous operations such as welding, spray painting, gluing, and carrying of heavy loads such as chassis and press plates. In addition, repetitive operations that require intricate motions such as assembly of odd components of printed circuits are very successful with robotics. These components might not be able to be loaded on PCBs using standard automation because of their special shape, weight, or lead pinout. They include relays, switches, lights, and other nonstandard electronic components.

Robotic assembly can be augmented with more sophisticated "intelligence," such as machine vision to select the part orientation for assembly. The robots can be integrated into a U-shaped line to perform different assembly functions, and they can also be integrated with more than one robot arm per assembly station to increase production rate. Robots can be reused for the assembly of the next generation of products, so that their salvage value is greater than that with hard automation.

The most important factor in the success of a robotic assembly process is the design of the parts so that they are easily presented to the robotic arms. Making parts as symmetrical as possible, or exaggerating their asymmetry with location guides and surfaces, is important. In addition, off-line loading of the parts in a suitable tray with a shuttle system for carrying trays in and out of the assembly stations is very successful. This type of system is used to assemble the Sony Walkman and the Polaroid camera shutter assemblies (see Section 8.3).

8.2 DESIGN RATINGS FOR MANUAL ASSEMBLY

The new methods of DFM have taken the design guidelines concepts of the early period and added a mathematical framework. These methods estimate the cost of assembly and quantify the importance of following the design guidelines through calculating penalty points for designs that are not ideal, either in part geometry or in assembly movement. The addition of these penalty points in reference to the ideal design or assembly motions forms a design efficiency rating. The methods are similar to historical time-and-motion study systems of measuring work through standard tables, such as methods time measurement (MTM), that industrial engineers use to estimate the time for assembly jobs on the production floor. The important difference is that these studies are done prior to completion of the product design, and assembly time can be reduced on the basis of design modifications in the early product development stages, resulting in more efficient assembly.

An important feature of these efficiency rating systems is that they can be used to evaluate competing designs or to improve current ones. They provide a common language for design and production engineers to communicate with each other on the applicability of the design to manufacturing issues and concerns.

In the United States, Professors Boothroyd and Dewhurst of the University of Rhode Island have been the research leaders, notably with the design of the IBM Proprinter in 1982. They have published their method in the book *Product Design for Assembly*. Their method is a system of rating each part in the assembly by assigning numbers to each on the basis of geometry, then adding up the numbers. The resultant number is a figure of merit for comparing the proposed design to an ideal theoretical one by calculating a design efficiency factor. The assembly is then redesigned using three simple guidelines to help in reducing the number of parts and the assembly time. The efficiency factor is then recalculated. The difference in the factors is the design savings. This method covers designs for manual, automatic, and flexible assembly.

Another method, developed by Hitachi in Japan, and adopted under license by General Electric in the United States, is the analysis of products or designs on the basis of assembly and joining motions. These motions are then given penalty points if they are not ideal, which is a downward assembly motion. The motions are added up to produce a normalized total that is compared to the ideal number of downward motions. The ratio of the two is the design assembly efficiency. There are no specific guidelines for increasing the efficiency number, but the rating system is adaptable to evaluation of competing designs.

Table 8.1 summarizes comparison of the two methods. Many companies have used and felt comfortable with either system in achieving good DFM results. Of the two systems, the Hitachi/GE method requires strong manufacturing knowledge in order to input the assembly sequence and motions. The Boothroyd–Dewhurst method is focused on part geometry, without any preference to the actual motions required for assembly. On the other hand, the importance of parts reduction in the Boothroyd–Dewhurst method tends to increase the use of snap fits and other nonfastening methods,

TABLE 8.1. Comparison of DFM Efficiency Techniques

Boothroyd/Dewhurst	Hitachi
Emphasis on part design	Emphasis on assembly process
Includes equipment costs for automation/robots assembly	Focuses on manual assembly
Compares to ideal design	Compares alternatives
Focuses on reducing part numbers	Does not penalize too many parts
Measures assembly time from given tables	Normalizes assembly time to standard downward motion

which could lead to design of assemblies that might be difficult to repair or disassemble.

Many comparisons have been made using both of these methods as well as MTM (methods time measurement) to calculate the design efficiency and the total assembly time for the same electronic product. The assembly time estimated by both methods usually is close, with many examples of assembly time calculations for household electronic products performed by University of Lowell graduate students being within 20%. When different groups estimated the assembly time of sample parts using the same efficiency method, they were within a similar range of 10–20%. For assembly efficiency, the numbers are quite different with each method, but the difference narrows to less than 20% when the ideal number of parts (as prescribed by Boothroyd–Dewhurst) is reached.

8.3 DESIGN FOR AUTOMATION AND ROBOTICS

As indicated by Figure 8.1, the cost of the automation or the investment in robotics is an important factor in the design of products for automation. The choice equipment, tools, and methods to be used, and the design of the automation line, have to be carried on concurrently with the design of the product. Many early automation attempts by major companies failed because products were selected for automation after their design had been completed. In determining the product design features that facilitate automation, the automation plan has to be well integrated with the product design and marketing plans, using the following guidelines:

- The forecast sales volume is reciprocally dependent on the automation plan. The plan will impact the production cost of the product and it will allow for some flexibility in determining the selling price and, indirectly, the expected sales volume. An automation strategy should be formulated that depends on the overall company and product strategy.

    In the successful implementation of an automation strategy, the expected profit level is determined initially on the basis of market dynamics studies. This level will determine the selling price and the planned production rate. Product design and process automation are then developed concurrently. This strategy is very common when electronic companies target a particular market segment and use automation as the competitive weapon. An example is the low-cost terminal developed in the mid-1980s by Hewlett-Packard Corporation in a concurrent product process effort in Roseville, California. Low-cost,

DESIGN FOR AUTOMATION AND ROBOTICS

traditional, and proven production technologies such as through-hole and single-sided PCBs were used to produce the terminal at a completely automated facility with specially designed component placement and inspection equipment connected together in an automated line.

- Determine the level of automation to be achieved. Although the goal of 100% automation is a very attractive one to set for the design team, the last 5% of the manufacturing process to be automated might require a much larger effort than can be justified. In addition, designing a fully automated production plant, and then reconsidering late in the cycle and partially substituting manual operations, is very difficult. In most cases, space and safety concerns needed for the manual operators have not been addressed in the initial robotic automation plans. It is also difficult to integrate both automatic machines and human operators in the same area and to share production control information at the same rate, humans being much slower than machines.

  It might be more beneficial to select specific subassemblies for automation, especially if they fit the traditional successful uses of automation. Examples are safety, health, and repetitive-motion assembly operations such as welding, spray painting, gluing, assembling heavy parts, and odd-shaped component assembly into PCBs.

- Design for transport of material, parts, and assemblies. In most automation plans, the focus is on the assembly operations: how to design parts for easier locating, grasping, and joining. The material movements from one station to the next is not emphasized. This is sometimes referred to as the "islands of automation" concept.

  An important part of the automation plan is the design for movement of the individual parts from the stockroom to the automatic lines and between stages of the same lines, including optimizing the number of trips versus planned storage space at the lines and in the stockroom. While there is wide availability of automation equipment, methods and technologies, transportation automation solutions are divided between conveyor lines and automatic guided vehicles (AGVs). The problems involved are the rate of material movement, the space needed to convey the material (AGVs are more efficient space wise than conveyors but are generally slower), and in-process inventory space needed for balancing the lines and the stockroom. Additional issues are the identification and decision-making techniques for the material conveyance and the disposition of spent material carriers and rejected assemblies. Integrating these problems will require skills and experiences that might not be readily available, especially when designing automatic lines for the first time. There are few automation suppliers who can solve both the automation and the transportation problems.

It is generally acknowledged that the use of computer simulation is the best technique for studying the transportation problem in automation. Software simulation packages are available, many able to run on personal computers, that are quite powerful and easy to use. Many simulations can be done with user-friendly interface menus, requiring no programming effort. The simulation results can be obtained in tabular, graphical, and animated forms. Simulation does not provide specific answers to a problem, but can be used to evaluate different alternatives. Each automation and material conveyance scenario has to be modeled; then reactions to specific inputs can be documented and compared in order to arrive at the optimum solution. Simulation models of the production operation should be initiated to study different alternatives in the transportation scenarios, and then kept current while the design of the system matures and also after production release. The simulation models could then be used as a basis for fine-tuning the automation system in production.

- Determine the automation system design. To expedite the automation system design and implementation, equipment should be purchased as much as possible from among existing automation suppliers and technologies. This will facilitate solution of support, maintenance, and capability problems and upgrades in the production stage. Design of company-specific automation devices and equipment should be avoided as much as possible, as it commits the production department to locking in the expertise required for running and maintaining the automation. Although it might be exciting to the company's team to design and implement specific automation, the support and maintenance phases of in-house automation are difficult to hand off to other engineering organizations, because the project might not fully meet its original goals or might be poorly documented. Many automation design teams tend to break up after the completion of the project because there is no immediate follow-on project and because the prospect of maintaining the current system is not as attractive as originally designing it.
- Design the parts for automation. The tenets of design for manufacture for manual assembly are valid for automation as well. Simplifying the part geometry for grasping and joining and minimizing assembly motions and part numbers are valid for robots as well as human operators. One important distinction is the emphasis on design for transportability. Design features added to aid the transportation of parts, such as addition of hooks or holes to surfaces for hanging individual parts, or alteration of the center of gravity to allow the parts to stand on one axis while being transported to the joining processes, are recommended. Vibratory motion, either in vibratory bowl feeders or in vibratory trays, where the parts are dumped on the trays then

DESIGN FOR AUTOMATION AND ROBOTICS

shaken into position, are the most common methods of presenting parts to the automation equipment. The parameters of the vibratory feed stations can be determined using the techniques of robust design discussed in Chapter 6. Parameters such as the frequency, stroke length, and duration of the vibration can be set for each axis of the vibration systems using the techniques of design of experiments.

In many cases, providing package for the parts that acts as both the storage element and the transport mechanism is very useful. The packaging sticks that integrated circuits are delivered in can be used in many different ways: as storage mechanisms; as protection from static, shock, and vibration; and as loading devices for automatic placement of components into PCBs. The same principles can be used for other parts in electronic products. The design of the package should be included in the task of designing the part.

- Integration of the automation into the plant-wide control system. The automation line should be designed to be well integrated with the plant production control and quality information systems. The automation system should have inspection stations throughout the operation to check the quality of the work performed, as well as an internal system to monitor the performance of individual automation stations. Data such as misfeeds, misloads, line and feeder jams, and the number of duration of line stops should be collected automatically. In addition, support information for automatic line defects such as times, locations, material supplier, material lot numbers, and date codes should be tabulated and made available. Control charts data to monitor the status of the line should be available on demand.
- Automatic line efficiency measurements. In addition to collecting data for the quality of the different automation stations, the most important measure of an automatic line is the actual rate of production versus the maximum equipment operational rates:

$$\text{Line efficiency (\%)} = \frac{\text{actual production rate}}{\text{theoretical maximum machine rate}} \times 100$$

This single number is very important to consider in terms of all the possible causes: parts misfeeds, jams, and incorrect material delivery; wrong and out-of-specification material; equipment breakdown; and poor operator training leading to misadjustment, incorrect setups, and unscheduled maintenance. These causes should be quantified and analyzed in a Pareto diagram, to prioritize attacking and fixing of the problems. Monitoring the automatic line for quality and performance problems also serves in the formulation of design guidelines for the next set of products to be made on this or similar automated lines.

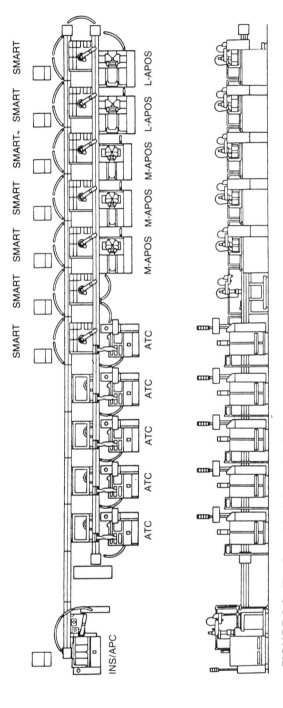

FIGURE 8.2. The Sony Multi Assembly Robotic Technology (SMART) robotic assembly line at Polaroid Corporation.

An Example of Design for Automation

An example of designing for automation is given by the camera shutter assembly made at the Polaroid Corporation assembly plant in Norwood, Massachusetts. The shutter assembly is a very complicated set of parts, springs, dials, and mechanisms that count the pictures used and set the shutter speed. The line used is a Sony Multi Assembly Robot Technology (SMART) robotic assembly line purchased from Sony Corporation. There are several vibratory feeding stations known as the Automatic Parts Orienting System or APOS. The APOS feeding stations load the parts from buckets into pallets, which are then dispatched to the assembly stations. Each pallet contains several cavities to hold the parts for transportation to the assembly stations and to form a feeding position for the robot arm. The robots will continue to load parts from the pallets in a sequential manner until a programmed number of cavities in a row are found to be empty.

The APOS stations are designed to dump the parts onto the pallets, vibrate the pallets to fill the cavities, retrieve the parts that have not been filled, and restart the loading operation a programmed amount of times. The line is designed so that all the APOS stations load in a predetermined rate to insure continuous operations.

An important measure of the line is the palletizing efficiency, which is the number of filled and properly oriented pallet cavities for a given part after

Figure 8.3. A view of the APOS loading stations.

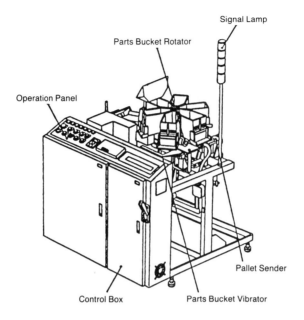

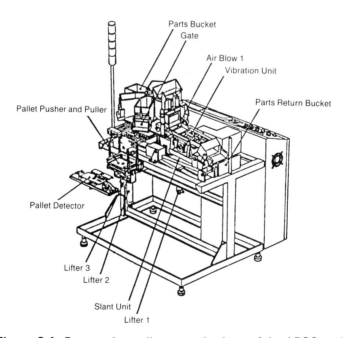

Figure 8.4. Front and rear diagrammatic views of the APOS station.

DESIGN FOR AUTOMATION AND ROBOTICS

the APOS cycle is completed, divided by the total number of available cavities. The success rate of the APOS loading system depends on two factors: the design geometry of the parts and the settings of the APOS machine parameters.

The APOS machine parameters are of more than twenty different types. They deal with the frequency, amplitude, and time for three axes of vibration as well as the angle of the pallet and the duration and frequency of air-blowing during the loading operation. The use of design of experiments and robust design techniques could be useful in determining the importance and the setting of each parameter to increase the palletizing efficiency.

The geometry of the parts to be assembled is also important in increasing the palletizing efficiency, as well as the ease of the joining motion for the robots. The same tenets of ease of handling and insertion of manual parts apply for the robotic assembly of parts. A symmetrical gear drive pick spring will have a greater efficiency rating than a complex-shaped plastic part such as the shutter spring.

The speed by which Polaroid engineers were able to set up the line and begin high volume production in a matter of weeks is encouraging. The main contributors to success were the purchase of the proven automation technology and the company's depth of experience in designing parts for hard automation; the redesign for flexible automation was minimal. Figures 8.2 through 8.6 are photographs and diagrams of the lines, APOS loading station, and the shutter assembly.

Figure 8.5. View of APOS machine and rear of robotic line.

Figure 8.6. Polaroid shutter assembly.

8.4 EXAMPLES OF DESIGN FOR MANUFACTURE EFFICIENCY

A typical first project for determination of design for manufacture efficiency is to take some assembly from an actual company design or competitive designs and rate the efficiency. In order to ensure success and to train the design team in design efficiency ratings, the following are recommended:

- Perform functional analysis of the product or assembly, using the axiomatic theory of design. It is important to define exactly the functionality of the product the team is dealing with to determine cost savings by eliminating unnecessary parts of assemblies.
- Analyze the assembly process and techniques for the product. It is important to agree beforehand how the product might be manufactured, both manually in low volumes and automatically in high volumes.
- Analyze the existing design based on manual and flexible (robotic) assembly, using sketches and diagrams for each assembly.
- Redesign the parts, using the techniques outlined in this book, justifying functional, material, and assembly changes. Sketches of the new parts should be made. In addition, estimate any changes in the tooling or fabrication strategy and the estimated costs of changes in strategy.

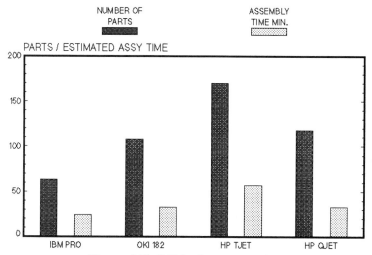

Figure 8.7. DFM printer comparison.

- List the assembly sequence in the manual and automatic cycles. Any tooling or fixturing needed to position the parts or subassemblies, as well as many assembly motions to get parts and subassemblies joined together, has to be noted.
- Develop a summary sheet outlining the redesign method and the dollar savings.
- Make a presentation to the design team management outlining the methodology and the benefits.

Figure 8.7 is an example of analysis of personal computer printers, comparing them to the IBM Proprinter in terms of both parts count and assembly efficiency. Other examples of parts that the design team could analyze for practice of these methods are common inexpensive household items such as hair curlers, fuzz removers, disposable film cameras, bathroom fixtures, video game switch controls, irons, or mechanical pencils. Table 8.2 shows a typical analysis performed by University of Lowell graduate students.

TABLE 8.2. DFM Assembly Analysis

Product	Original assembly cost ($)	New assembly cost ($)
Fuzz remover	30.56	27.44
Disposable camera	34.86	25.31
Video monitor	87.12	58.32
Smoke detector	211.75	178.54
	(circa 1980)	(circa 1988)

8.5 CONCLUSION

The design efficiency rating methods are important in quantifying new product designs for manual or automatic assembly. Several methods have been illustrated, and they are used with good success by different world class manufacturing companies. In setting the expected design efficiency, the new product team should examine the company's past products as well as the competition's, to determine the current level of design efficiency, and to set the expected level for new products.

Designing for automation and robotics is increasingly important as the cost and the degree of integration of robotic equipment becomes more affordable at lower volumes. The problems inherent in designing for robotics are similar to those for manual design, with the additional complexity of how to present the parts to the robot grippers. The example of the Polaroid Corporation robotic line illustrated some of these problems.

SUGGESTED READING

Andreasen, M. and Hein, L. *Integrated Product Development.* Kempston: IFS Publications Ltd., 1989.

Andreasen, M., Kahler, S. and Lund, T. *Design for Assembly.* Kempston: IFS Publications Ltd., 1983.

Anon. "Designing parts for automatic assembly." *Manufacturing Engineering,* pp. 46–50, July 1980.

Asfahl, C. Ray. *Robots and Manufacturing Automation.* New York: Wiley, 1986.

Boothroyd, G. and Dewhurst, P. *Product Design for Assembly.* Wakefield, R.I., 1987.

Bralla, J. (ed.). *Handbook of Product Design for Manufacture.* Dearborn, Mich.: Society of Manufacturing Engineers (SME) Press, 1988.

Eder, W.E. "Information systems for designers." Proceedings of the Institution of Mechanical Engineers, ICED 1989, Harrogate, UK.

Groover, Mikell P. *Automation, Production Systems, and Computer Aided Manufacturing.* Engelwood Cliffs, N.J.: Prentice-Hall, 1980.

Hubka, V. "Design for quality." Proceedings of the Institution of Mechanical Engineers, ICED 1989, Harrogate, UK, pp. 1321–1333.

Hurthwaite, B. "Designing in quality." *Quality Journal,* pp. 34–40, Nov. 1988.

Lane, Jack D. (ed.). *Automated Assembly.* Dearborn, Mich.: Society of Manufacturing Engineers (SME) Press, 1986.

Leverett, T. *The Architecture of a Small Automated Assembly Line.* Charlotte, N.C.: IBM Corporation.

Lotter, Bruno. "Using the ABC analysis in design for assembly." *Assembly Automation,* pp. 80–86, May 1984.

Nanji, Bartholomew. *Computer Aided, Design Selection and Evaluation of Robots.* New York: Elsevier, 1987.

Nazemetz et al. *Computer Integrated Manufacturing Systems, Selected Readings.* Institute of Industrial Engineering, 1986.

Poli, C., Graves, T. and Gropetti, R. "Rating products for ease of assembly." *Machine Design*, August 1986.

Shina, S. "Developing a course in design for manufacture." *Journal of Industrial Technology*, Vol. 7, No. 3, 1991.

Waterebury, Robert. "Designing parts for automated assembly." *Assembly Engineering*, pp. 24–26, Feb. 1985.

Zurn, J. "Assessing new product introduction using the problem discovery function." *Quality Engineering*, Vol. 2, No. 4, pp. 391–410, 1990.

CHAPTER 9

Geometric Dimensioning and Tolerance Analysis

This chapter is divided into two parts: the first discusses the subject of geometric dimensioning and tolerancing and its relevance in design for manufacturability. The second part deals with the why's, how's, and the need for tolerance analysis in taking a product from design to manufacture.

In this era of multinational corporations it is difficult not to visualize multinational or international manufacturing. The decade of the 80s witnessed many U.S. corporations going abroad to manufacture their products. Likewise many foreign companies set up manufacturing facilities in the United States and are turning out products for the U.S. and foreign markets. Most notable examples of these are Japanese automobile corporations like Honda, Mazda, and Toyota. This trend will not only continue but will increase in the foreseeable future. The market forces of demand and supply as well as competition compel organizations to seek the best resources for their products. Economies of scale, availability of skills, labor, and materials as well as specialization will force companies to find the best match for producing the goods.

Products designed in one country have components that are made and purchased from other countries, and these are assembled in a third country. In all cases the parts are expected to mate with their counterparts and function as per the design intent. The product designers, engineers, and manufacturers need a standard system that can be read and interpreted by technical people all over the world. The ANSI (American National Standards Institute) Y14.5M-1982 standard has been developed primarily to address these issues. New symbols and concepts have been added to the old dimensioning and the plus and minus system of tolerancing. These symbols exercise control both in the way the part is manufactured as well as in the way it is inspected.

The result is a system that not only offers flexibility to the designer in terms of design but also directs the way the part is manufactured. Geometric dimensioning and tolerancing (GDT) is a methodology that persuades the designer to evaluate objectively from both the functional and manufacturing perspectives. Since this is done in the early phase of project design, GDT has often been referred to as a design for manufacturability tool. Proven tangible benefits of GDT so far have been repeatability, part cost reduction, expanded tolerances, better communication because of standardized design language, and clarity of ideas conveyed.

An in-depth study of GDT methodology and its applications is beyond the scope of this book. This chapter's objective is to acquaint and impress the reader with the importance of GDT as a design for manufacturability tool and its usefulness in today's competitive marketplace.

Tolerance analysis involves the analysing of parts with respect to assemblability and, more importantly, functionality. The significance of tolerance analysis can never be underestimated. There is a definite dimensional spread change from parts that are made and used for prototypes to those for production. Design has been divided into two completely different categories: design for prototypes and the second for production. The objective for each phase is different and the processes—manufacturing, assembly, and usage—are all different. Failure to comprehend this means that the risk of failure is very high. Almost everyone can recall at least one incident when a product worked flawlessly in the prototype phase and turned into a debacle when released to production. Tolerance analysis is a tool that can be used to avoid such situations.

9.1 GDT ELEMENTS AND DEFINITIONS

Figure 9.1 illustrates the basic GDT elements. The front and side views of the part have been dimensioned and toleranced using GDT methodology:

- A, B, C, and D are datums derived from datum features. Datum features are surfaces of a part used to establish perfect planes. These planes are used as origins of measurements for the hole locations, squareness, perpendicularity, angularity, and so on.
- There are two different sizes of holes, which are indicated by the diameter symbol along with the linear tolerances. (Dia. 0.375–0.380) and (Dia. 0.750–0.760).
- Basic dimensions such as 1.000. These dimensions are not toleranced. Their tolerances are accounted for in the feature control frame.
- *Feature control frames* consisting of the following:

  A geometric characteristic symbol—in this case two location controls
  Zone descriptors—in this case diameters (0.375 and 0.750)

210 GEOMETRIC DIMENSIONING AND TOLERANCE ANALYSIS

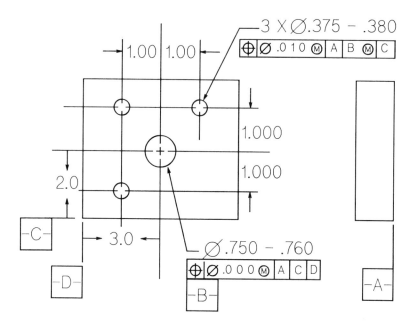

Figure 9.1. Basic GDT elements.

Tolerance of location—for example, Dia. 0.010 @ M (maximum material condition)
A material condition symbol—M
A, B, and C are the primary, secondary, and tertiary datums.

GDT comprises a group of concepts and symbols that serve to convey the design concept and intent to the fabricator and inspector. The objective is to constrain the manufacture and measurement of a part to one method so that repeatability and consistency are assured.

GDT is the result of many years of study conducted by ANSI, with international inputs, to produce a universal design language. GDT is gaining worldwide acceptance and usage because of the increase in international manufacturing. GDT elements are broadly divided into five categories:

Cylindrical tolerance zones
Datums
MMC, LMC, and RFS
Controls: form, orientation, profile, runout, and location
Feature control frames

9.2 CYLINDRICAL TOLERANCE ZONES
(See Figures 9.2 and 9.3)

The most convenient and popular method of *locating* features (holes, slots, machined steps) up to now has been the plus and minus system for location tolerancing. This has also been referred to as *true positioning*.

The part in Figure 9.2 is assigned a plus or minus tolerancing system for locating the hole from the edges. The dimension in the X and Y axes from the corresponding surfaces (edges) is 2.000 ± 0.005 inches and 3.000 ± 0.005 inches. This means that the center of the hole in the X axis can be located from the related surface within the limits of 1.995 and 2.005 inches. The location for the center of the hole in Y axis are 2.995 and 3.005 inches.

Plotting the center location of the hole in both the X and Y axes produces a square tolerance zone (see Figure 9.4) of 0.010×0.010 inch in which the center of the hole can be located. The four points on the corners of the square and the center of the hole can be 0.007 inch from the true center, while other points are 0.005 inch from the true center. This creates a dilemma that is difficult to explain—why should there be only four points 0.007 inch away from the center whereas other points are 0.005 inch from the center? For computing mating conditions, the worst case or the maximum out-of-position

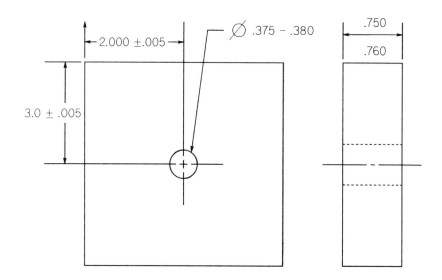

Figure 9.2. Plus or minus tolerancing.

212 GEOMETRIC DIMENSIONING AND TOLERANCE ANALYSIS

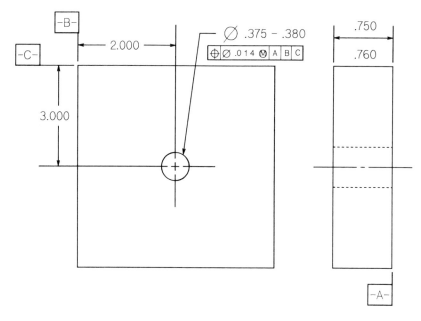

Figure 9.3. Cylindrical tolerancing.

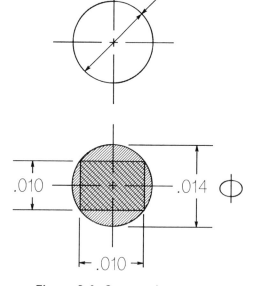

Figure 9.4. Square tolerance zone.

tolerance is taken into account: in this case 0.007 inch away from the true center.

Hence the GDT methodology has moved away from the square tolerance zone to a cylindrical tolerance zone of 0.014 inch Dia, wherein all the points for the center of the hole are positioned. This adds 57% more tolerance without jeopardizing the mating condition or the function of the part. As seen in Figure 9.3, there is also an M (maximum material condition) control attached to it. This means that the cylindrical tolerance zone of 0.014 inch Dia applies at the hole size of 0.375 inch diameter.

It can be shown that as the hole grows larger (from 0.375 to 0.380 inch) and the corresponding mating part (for example, a shaft) is made to the smaller side of the tolerance, the overall tolerance zone gets larger. So we can derive the benefits of having a larger tolerance zone for the parts without any danger of the parts not mating.

The GDT system only recognizes the cylindrical tolerance zones. Its use is becoming universal. The plus and minus system will always be used for specifying holes, slots, and feature sizes. It is definitely advantageous to use the cylindrical tolerance zones for positional tolerancing.

9.3 DATUMS

The concept of datums is one of the most important characteristics of GDT methodology. Datums have been used since the inception of technical drawings for highlighting certain areas in a drawing for perpendicularity, parallelism, and so on. The GDT system has structured the use of datums to tie the part down. This means that once the datums are specified then the manufacturing process is also defined. Hence the latitude for the manufacturer to make this part by different processes is reduced. The result is that a part made by different manufacturers will meet the design specifications.

Datums are points, lines, or planes that are used as bases from which the geometry of the part is defined and controls are applied (see Figure 9.6). They can be on the part or outside the part. Since they are perfect, they do not exist in reality and can only be simulated. Well-defined and permanent features of a part are used to serve as datums. These are appropriately called *datum features*.

To understand the application of datums, let us take an example of a free-floating body that has to be machined (see Figure 9.5). Any free-floating body has six degrees of freedom—three in the linear axes of X, Y, and Z and the other three in the rotational axes. A part to be fabricated has to be tied down. This is normally done by clamping the part on the machine tables

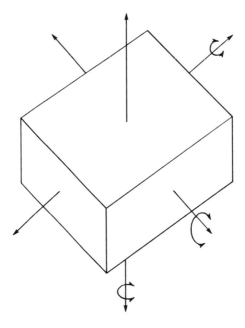

Figure 9.5. Datum reference frame.

in vises, magnetic chucks, and the like. After the operation is performed the part is released and is prepared for another operation and the process is repeated. Preparing a part for performing any operation is normally called a *setup*.

To perform any operation such as milling, cutting, drilling, boring, or painting the location has to be specified. The location is related to and measured from a base or starting point that has some coordinates in space, for example (0, 0, 0). The feature locations for the rest of the part are measured from this base (with (0, 0, 0) coordinates). The base, which could be a point, a line, or a surface is called a *datum*: in other words, something to measure or calculate from.

A datum frame basically consists of three mutually perpendicular planes (see Figure 9.6). The objective is to lock the part in the three planes so that it relinquishes all its degrees of freedom: in other words, it has zero degrees of freedom. A minimum of three high points are needed to establish the part in the primary datum plane. Once these are established and the part makes contact with all three points, two rotational degrees and one linear degree of freedom are eliminated. A minimum of two points are required to locate and tie the part in the secondary datum plane (see Figure 9.6) while

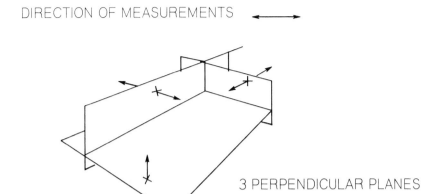

Figure 9.6. Free-floating body representation.

maintaining the three-point contact in the primary datum plane. This eliminates two more degrees of freedom, one linear and one rotational. Finally, one high point is needed to hold the part in the tertiary plane. This eliminates the final linear degree of freedom, providing a positive part orientation for any manufacturing or inspection operation.

Since datums are defined as perfect and not existing in reality, the primary, secondary, and tertiary datum planes used above to tie the part down can only be simulated. Machine tables, chucks, surface plates, angle plates, Vee blocks, and the like are all examples of equipment used to simulate datums. It is not always necessary to specify all three datum planes. In many cases only two or sometimes one datum plane is required to tie a part down. In such situations, specifying additional datum planes is unnecessary and becomes expensive.

Datum features are used to simulate datums. They are well-defined permanent features and are selected on the basis of geometric relationship to the toleranced feature and the design requirement. The mating features of other parts are also selected as datum features. This ensures proper mating at assembly. It is extremely important to select datums carefully. The primary, secondary, and tertiary datums are assigned keeping part functionality, mating conditions, and manufacturability in mind. Switching datums downstream extracts a heavy penalty by imparting a large difference in part orientation that can affect part functionality and its assembly with other parts.

There are many types of datums:

Permanent datums
Temporary datums

Compound datums
Partial datums
Cylindrical datums

Permanent datums, as the name suggests, are datum features that are not changed or altered by any subsequent machining or forming operations. *Temporary datums* are normally specified on castings, for example, for machining of permanent datums. Once the machining operation is completed, the temporary datums cease to exist. *Compound datum* features comprise two datum features used to establish a single datum. These are commonly used in depicting situations of coplanarity and coaxiality. In cases where parts are large and do not require using the whole surface as a datum, a small portion of the surface is used to serve as a datum. This surface is designated as a *partial datum.* Partial datums are commonly used in large weldments, castings, and where the functionality is limited to a small portion of the part.

It is also common to choose cylindrical features as datums. A *cylindrical datum* feature is associated with two theoretical planes intersecting at right angles on the datum axis. Once the datum axis is established, it becomes the origin for related dimensions, while the two planes (X and Y) indicate the direction of measurement.

9.4 MMC, LMC, AND RFS

Three new and important concepts that have been added by the GDT methodology are maximum material condition, least material condition, and regardless of feature size.

Maximum material condition (MMC) is the situation in which a feature of size (such as of a hole or shaft) has the maximum amount of material within the stated limits. For example, a hole that has a diameter and tolerance of 0.500 to 0.510 inch: at the maximum material condition, the diameter of the hole will measure 0.500 inch. Similarly, a shaft with a diameter and tolerance of 0.500 to 0.510 inch has a diameter of 0.510 inch at its maximum material condition.

What this means is that at MMC the *part* (not the hole or the shaft) has the maximum amount of material when weighed or measured physically. This explains why at MMC, the shaft has the maximum diameter and the hole has the minimum diameter.

Least material condition is the opposite to MMC. In this situation a feature of size (such as a hole or shaft) has the least amount of material within the stated limits. For example, a hole with a diameter and tolerance of 0.500

to 0.510 inch has 0.510 inch diameter at the least material condition. By the same reasoning, the shaft with a diameter and tolerance of 0.500 to 0.510 inch will have a diameter of 0.500 inch at the least material condition.

Regardless of feature size (RFS) is the situation in which controls are applied independently of the size of the feature within the stated limits. This is normally used when specifying the orientation controls of perpendicularity, parallelism, and angularity. When a geometric tolerance is applied on an RFS basis, the tolerance is limited to a specified value regardless of the actual value of the feature. Therefore, if a datum feature is referenced on an RFS basis, a centering about its axis or center plane is necessary regardless of the actual size of the feature.

9.5 CONTROLS

Controls are symbols designating the variety of conditions that a feature has to meet to achieve a desired shape or form (Table 9.1). These could be perpendicularity, parallelism, angularity, or controls for profile. The shape or form is of course dictated by the design. Along with the control symbols, tolerances are specified so that the features can be produced within limits. There are five types of controls: form, orientation, profile, runout, location.

TABLE 9.1. Standard Geometric Characteristic Symbols

	TOLERANCE TYPE	CHARACTERISITIC	SYMBOL
INDIVIDUAL FEATURES	FORM	STRAIGHTNESS	—
		FLATNESS	▱
		CIRCULARITY (ROUNDNESS)	○
		CYLINDRICITY	⌭
INDIVIDUAL OR RELATED	PROFILE	PROFILE OF A LINE	⌒
		PROFILE OF A SURFACE	⌓
RELATED FEATURES	ORIENTATION	ANGULARITY	∠
		PERPENDICULARITY	⊥
		PARALLELISM	//
	LOCATION	POSITION	⌖
		CONCENTRICITY	◎
	RUNOUT	CIRCULAR RUNOUT	↗
		TOTAL RUNOUT	↗↗

FORM The controls for straightness, flatness, circularity and cylindricity all control the form a feature can take. These controls are individual (unrelated). Unrelated features are associated with the control of a feature, without consideration of other features.

The reasoning for this becomes very apparent when in the figure specifying flatness there is no relationship to any other features. Features that are compared to their theoretical perfect planes, axes, or surfaces are termed *individual* or *unrelated controls*. When a feature is selected for use as a datum, it is important to specify flatness as a means of controlling that feature. GDT very clearly defines form controls of straightness, cylindricity, and so on. The form controls are applied by means of a tolerance band or zone, within which the surface or axes must lie (Figure 9.7).

ORIENTATION (See Figure 9.8.) Angularity, parallelism, perpendicularity, and in some cases profile are classified as *orientation controls*. These are related features that are the opposite of individual features. The tolerances control the orientation of features to one another. They are sometimes referred to as *attitude tolerances*.

The tolerance is normally defined by a tolerance zone (or band) within which the axis or the surface must lie (Figure 9.8).

PROFILE Profile is used to control form or a combination of size, form, and orientation. Profiles are of two types: profiles controlling lines and profiles controlling surfaces. Tolerance zones are used to control the forms. Where profiles or surfaces are to be controlled, control is easier to apply if the surfaces have a defined cross-section.

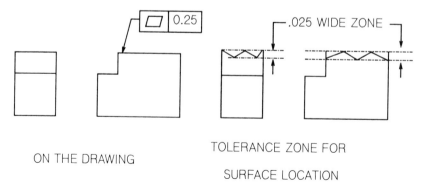

ON THE DRAWING

TOLERANCE ZONE FOR SURFACE LOCATION

Figure 9.7. Specifying surface flatness.

FEATURE CONTROL FRAME 219

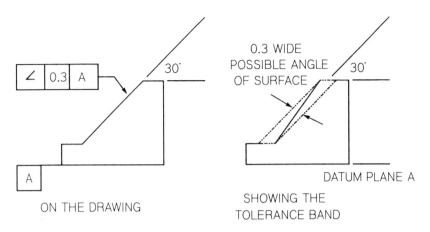

Figure 9.8. Specifying angularity.

RUNOUT Runout is a *composite tolerance*. It is used to control the functional relationship of one or more features of a part or a datum axis. Runout controls are of two types: circular or total. Each is applied depending on the design requirements.

LOCATION Location has been discussed earlier; it is used when comparing the plus or minus system of tolerancing to the cylindrical tolerance zones. Location control is essential to the functional requirement of the part.

9.6 FEATURE CONTROL FRAME (See Figure 9.9)

The positional tolerance, feature information including the datums, and their relationships are contained in a feature control frame.

A *feature control frame* consists of a box divided into compartments containing the geometric characteristic symbol followed by the tolerance. Conditional symbols such as M (maximum material condition), L (least material condition), and R (regardless of feature size) are also included.

The amount of information is dependent on the design requirement and individual geometric characteristics. When a geometric tolerance is related to a datum, the relationship is indicated by a datum letter in the compartment following the tolerance. When a feature is associated with more than one datum (primary and secondary), the primary datum is referenced first by placing it next left from the tolerance. Feature control frames can be combined with datums, projected tolerance zones, and so on for additional information.

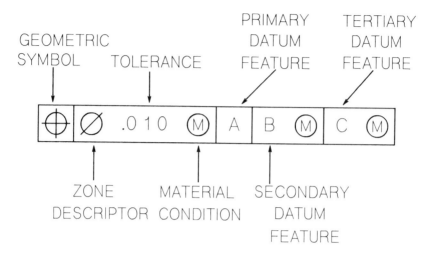

Figure 9.9. Feature control frame.

9.7 TOLERANCE ANALYSIS

The tolerance analysis concept is not new. Ever since parts have been fabricated and assembled to function as products some form of tolerance analysis has been done. Before the advent of the industrial era the products were simple, comprising few parts, and all these parts were made in one shop. The use of materials was also limited to wood and metal and the manufacturing processes were limited to forging, casting, and some elementary machining. Parts were designed to nominal dimensions and some tolerances were specified on the basis of knowledge of the process. If one dimension was made out of specification, the mating part was modified to correspond to the specification of the first part. The volume was low, and labor was comparatively inexpensive.

With the industrial era came mass production and the processes of design and manufacture of products changed completely. New materials and processes came into being and the luxury of a couple of people making a product in one place ceased to exist. It was important to manufacture the parts within the specified limits. New and complex machine tools were available to make complex shapes and sizes. New processes such as sheet metal fabrication on a mass scale were being used; new materials like rubber and plastics were also being put to use. Hence tolerance analysis becomes a necessity.

9.7.1 Product Design Process

The product design process in most companies starts with the concept models and moves through prototype models to production. The concept and prototype models are primarily made in the company's model shop or outside machine shops where most of the individual components are fabricated by one or two tool makers. The emphasis here is to prove the concepts. The tool makers work with the designer and are given some latitude in making the parts to fit. The product is still being designed at this stage, so changes are frequent and the parts are altered to fit. Because of time constraints, the changes are drawn as freehand sketches and given to the tool makers.

When the models are completed and assembled they go through extensive testing. More changes are made and incorporated. After testing is completed, drawings are updated to reflect the changes. Suppliers are selected and orders are issued to produce parts.

When parts are received for the first time and are assembled for production, it is frequently discovered that they do not fit. At this time it is also discovered that, due to schedule pressures, detailed tolerance analysis was not done. Drawings made and released based on the concept and prototype models did not account for the manufacturing process capabilities.

Situations such as these not only happen but are repeated, every time a new product is designed and implemented in production. This process of new product design should be divided into two separate categories: design for prototypes and design for production. Each of these has different objectives and should be clearly understood.

- *Design for prototypes.* The main goal is to prove the concept for functionality. Design for manufacturability is not high on the list of priorities. Parts are made in the company's model shop or one of the machine shops, where only one or two tool makers are assigned to work on the parts. Changes are being incorporated as the parts are being made. If parts do not fit at assembly, they are modified for fit and function.
- *Design for production.* Parts are made in quantity by different suppliers. Extensive tolerance analysis is done with knowledge of manufacturing processes to make sure parts fit at assembly. Different manufacturing processes are used for producing the parts.

9.7.2 Why Tolerance Analysis?

It is universally accepted that there is no such thing as a perfect part, that is, a part made to exact dimensions. Applying this fact to the manufacture of parts, we can safely say that no manufacturing process can make a part to exact dimensions. Hence maximum and minimum limits of dimensions (or tolerances) are specified with two goals in mind:

- The limits must be set close enough to allow functioning of the assembled parts (including interchangeable parts).
- The limits must be set as wide as functionally possible, because tighter limits call for expensive processes or process sequences.

Once the limits (or tolerances) are set by the designer then all parts or components are manufactured to those specified limits. Assembly of the parts causes tolerances to accumulate, which can adversely affect the way the final product functions. Furthermore, tolerance accumulation can also occur in the way the part is dimensioned. Tolerance accumulation that occurs in the assembly of parts is sometimes referred to as *tolerance stackup*.

To make sure that parts successfully mate at subassembly or final assembly, and that products function as per the design intent, an analysis is performed to uncover the existence of any interferences. This study is formally referred to as *Tolerance analysis*. A review of the basics will aid understanding of what goes into tolerance analysis.

Tolerance (per ANSI Y14.5M-1982) is the total amount by which a specific dimension is allowed to vary. Geometric tolerance is a general term applied to the category of tolerances used to control form, profile, orientation, location, runout, and so on. Tolerances are primarily of two types: tolerance of size, and tolerance of form, location, ...

Tolerance of size is stated in two different ways:

- *Plus or minus tolerancing*, which is further subdivided into bilateral and unilateral tolerancing. Bilateral tolerance is applied to both sides of a basic or nominal dimension. For example,

$$0.375 \pm 0.010$$
$$0.375 + 0.005/-0.002$$

- *Limit dimensioning* is a variation from the plus or minus system. It states actual size boundaries for the specific dimension. This eliminates any calculation on the part of the manufacturer. Limit dimensioning is practiced in two ways—linear or one next to another, and dimensions placed one above the other. For example,

$$0.635 - 0.635$$
$$0.635$$
$$0.625$$

TOLERANCE ANALYSIS

When one dimension is placed above the other, the normal rule is to place the larger dimension above the smaller.

There are no hard and fast rules regarding the tolerancing techniques. The choice depends on the style of the designer and very often one finds both types of tolerancing methods (the plus or minus and limit dimensioning) used on the same drawing.

Tolerance of form covers location of geometric features and geometric properties like concentricity, runout, and straightness. These tolerances have been discussed earlier in the section on GDT.

9.7.3 Types of Tolerance Analysis

There are two types of tolerance analysis: extreme-case tolerance analysis, and statistical analysis. Before looking in detail at these two types of analysis, it is important to understand that the parts used in a product are divided into *standard* or off-the-shelf parts and *nonstandard* or designed parts.

Examples of standard parts are bearings, shafts, pulleys, belts, screws and nuts, and snap rings. These parts come with the manufacturers' specified tolerances, so the designer does not have any latitude in changing these limits. Nonstandard or designed parts are custom-made for the product. Hence the designer can specify wider or narrower limits based on the functionality requirements.

Extreme-case analysis is further subdivided into best-case analysis, and worst-case analysis (see the example in the statistical analysis).

- Best-case analysis describes a situation in which the parts fall on that side of the tolerance (positive or negative) in which there is no chance of interference in the assembly of these parts.
- Worst-case analysis is the study of the worst situation when the parts are assembled: the probability of interference is unity.

The extreme case analysis method is currently the most widely used and most popular. Most designs are based on this concept and have worked successfully through the decades. The method is simple to apply and consists of designing the parts to nominal dimensions and then assigning tolerances in such a way that if tolerances accumulate in one direction or the other, the assembly still meets the functional requirements of the product. This method, even though ensuring that all parts will assemble, has a built-in waste mechanism. Designs can be overly conservative, leading to high product costs by assignment of tighter tolerance zones. As we will see by using statistical analysis, we can have a better understanding of how and how many parts will assemble in a given set of conditions.

9.7.4 Statistical Tolerance Analysis

When several parts are to be assembled, the task of computing the total deviation of all the parts at their maximum limits can be so great that in practice it is not feasible. In situations such as these statistical analysis is increasingly being employed.

Statistical analysis involves the application of statistical probability distributions to analysis of tolerances for assemblys. It will prevent overdesign. Overdesign, or more specifically overly conservative design, increases the cost of the product without adding value and hence results in waste. With statistical analysis, tolerances can be widened and, if process capabilities are known, even wider tolerances can be assigned.

Tolerance analysis using statistics is based on the theory that most mechanical parts are made to normal probability distributions within their specified tolerance limits. The distributions of individual parts can be combined into one normal distribution representing the assembly deviation from its nominal dimension. In computing the mean and standard deviation of an assembly, the nominal dimension of a part is taken to be its mean and the specified tolerance limits of that part are taken to be the limits of the normal curve ($\pm 3.8\sigma$). For practical purposes $\pm 3\sigma$ is sufficient.

Let us take an example of three parts (for simplicity consider them to be rectangular blocks) that are to be assembled together, with the subassembly to be finally assembled into a box (see Figure 9.10). Their critical dimensions (mating surfaces) and their specified tolerances are as shown. What tolerance should be assigned to the 3.875 inch dimension in the box for these parts to fit?

The problem will be solved using extreme case analysis and then by statistical analysis. The results and logic are somewhat different.

9.7.5 Extreme-Case Analysis Example

Using the worst-case analysis, we see from Table 9.2 that the cumulative dimension of the three parts can be as high as 3.902 inches. Hence the minimum dimension of the box should be 3.903 inches to allow a 0.001 inch clearance between the box frame and three parts. Assigning a ± 0.005 inch tolerance, the mean dimension for the box is 3.093 + 0.005 = 3.908 inches. Hence:

Maximum gap: 3.913 − 3.848 = 0.055 inch
Minimum gap: 3.903 − 3.902 = 0.001 inch
Mean gap: 3.908 − 3.875 = 0.033 inch

Having such a wide variation (0.055 to 0.001 inch), may or may not be acceptable from a functional requirement. If this assembly, for example, were part of the front panel, a gap of 0.055 inch might be esthetically unpleasant and would convey the impression of poor quality.

TOLERANCE ANALYSIS

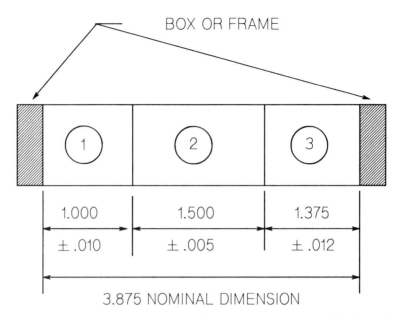

Figure 9.10. Tolerance analysis example, three square parts. (All dimensions in inches)

To reduce the variation, the next approach is to tighten the tolerances. Table 9.3 below gives the result of having the parts made to closer tolerances: assume all parts are made to ± 0.002 inch (see Table 9.3). In this case the minimum dimension of the box is 3.882 inches. If this dimension is assigned a ± 0.002 tolerance then, the nominal dimension becomes 3.884 inches, with a maximum of 3.886 inches.

Maximum gap: $3.886 - 3.869 = 0.017$ inch
Minimum gap: $3.882 - 3.881 = 0.001$ inch
Average gap: $3.884 - 3.875 = 0.009$ inch

TABLE 9.2. Tolerance Analysis Example—Normal Tolerance

Part	Nominal dimension (in)	Tolerance (in)	
		Dimension high	Dimension low
P1	1.000	1.010	0.990
P2	1.500	1.505	1.495
P3	1.375	1.387	1.363
Total	3.875	3.902	3.848

TABLE 9.3. Tolerance Analysis Example—Tight Tolerance

Part	Nominal dimension (in)	Tolerance (in)	
		Dimension high	Dimension low
P1	1.000	1.002	0.998
P2	1.500	1.502	1.498
P3	1.375	1.377	1.373
Total	3.875	3.881	3.869

This is more acceptable than the case with normal tolerances above, but it comes at a very high cost. The fact that we are narrowing the tolerance band basically means that we are not using the normal manufacturing process. It requires extra time in setup, increased inspection, and increased defect rate because of parts falling out of tolerances. It is a good example of an overly conservative design.

9.7.6 Statistical Analysis Example

With statistical analysis we assume that all the three parts are made to a normal probability distribution within their specified tolerance limits. For practical feasibility computation will be to $\pm 3\sigma$ (see Table 9.4).

Now the formula for the average (combined) standard deviation of the three parts is

$$\sigma_{avg} = \sqrt{(\sigma_1^2 + \sigma_2^2 + \sigma_3^2)} \quad (9.1)$$
$$= \sqrt{[(0.0033)^2 + (0.0016)^2 + (0.0040)^2]}$$
$$= 0.0054 \text{ inch}$$
$$\sigma_{avg} = 0.0054$$

We can sum up our analysis as follows:

$$\pm 1\sigma = \pm 0.0054 \text{ (or 0.006) inch}$$
$$\pm 2\sigma = \pm 0.0108 \text{ (or 0.011) inch}$$
$$\pm 3\sigma = \pm 0.0162 \text{ (or 0.016) inch}$$

TABLE 9.4. Statistical Tolerance Analysis Example

Part	Total tolerance (in)	3σ	$1\sigma = 3\sigma/3$
P1	T1 = ±0.010	0.010	0.0033
P2	T2 = ±0.005	0.005	0.0016
P3	T3 = ±0.012	0.012	0.0040

TOLERANCE ANALYSIS

The mean is the cumulative nominal dimension and is equal to 3.875 inches. From the normal tables in an earlier chapter we see that:

$$\pm 1\sigma = 68.3\%$$
$$\pm 2\sigma = 95.5\%$$
$$\pm 3\sigma = 99.7\%$$

INTERPRETATION If the dimension of the box was 3.881 (3.875 + 0.006) inches for ± 1 standard deviations, then 68.3% of the parts fabricated would fit.

If the dimension of the box was 3.886 (3.875 + 0.011) inches for ± 2 standard deviations, then 95.5% of the parts fabricated would fit.

If the dimension of the box was 3.892 (3.875 + 0.016) inches for ± 3 standard deviations, then 99.7% of the parts would fit. This means that two parts out of one thousand part combinations would have an interference fit and would require shuffling through the box of parts for a different combination.

The advantage of this system is that we can use the parts produced at standard specified tolerances and can *predict* the number of parts that will fit.

Note that there is an assumption that no interrelationships exists between the three parts and the box. Interrelationship exists when the production of one part influences the dimensions obtained in the other. If interrelationships do exist, then the formula used and the results obtained will be different.

9.7.7 Process Capability and Statistical Analysis

If the manufacturing process capabilities of the parts produced (P1, P2, and P3) are known, this offers us more flexibility in assigning tolerances.

For example, assume the statistical process control (SPC) data for parts P1, P2, and P3 are known, and assume that the parts are being produced within the control limits of ± 0.005 inch instead of the specified limits shown in Table 9.4. Then ± 0.005 can be used in our computations of average and standard deviations of the assembly, which will result in a much tighter distribution (see Table 9.5).

TABLE 9.5. Statistical Tolerance Analysis from SPC Data

Part	Total tolerance (in)	3σ	$1\sigma = 3\sigma/3$
P1	T1 = ± 0.005	0.005	0.0016
P2	T2 = ± 0.005	0.005	0.0016
P3	T3 = ± 0.005	0.005	0.0016

Putting these values into equation (9.1) yields the average standard deviation of ±0.0027 inch. This is equal to half the deviation that was obtained by using the specified tolerance limits for the parts, which clearly shows that knowledge of the process(es) is very important for a good tolerance analysis.

9.7.8 Tolerance Analysis and CAD

There is a proliferation of CAD (computer-aided design) systems these days, and with good reason. CAD systems make it convienient to make drawings in three dimensions and hence offer a better view of parts going together at assembly. Parts are drawn to nominal dimensions and can be checked for interferences. Parts can also be drawn at extreme-case tolerances and checked for interferences.

But one thing the reader should note is that CAD systems cannot be used for tolerance analysis. Tolerance analysis involves understanding of the functionality of the product, knowledge of the processes that are used in making the parts, and so on. Hence even if CAD has been used for making the prototypes, a separate tolerance analysis study should be done.

9.7.9 Tolerance Analysis and Manufacturing Processes

A product is made up of many parts that have been made from different materials. Many electronic products use parts made from plastics, sheet metal, machined parts, rubber, castings, and so on. Not only are parts made from these processes only unique in their properties, but the manufacturing dictates to a large extent the tolerances that can be specified on the parts. The sheet metal parts require a much wider tolerance band compared with machined parts. The plastic molded parts, once the molds have been fabricated and the process has been defined, tend to maintain a very low variation in dimensions from batch to batch. Machined parts will vary from batch to batch.

Knowledge of manufacturing processes will protect one from overly conservative design, which increases the cost of the product.

9.8 TOLERANCE ANALYSIS CASE STUDY

PROBLEM STATEMENT A hot stamping press is used to print on plastic film, as shown in Figure 9.11.

- Various widths of plastic films are used for different jobs.
- The stamping dies are changed from job to job.

TOLERANCE ANALYSIS CASE STUDY

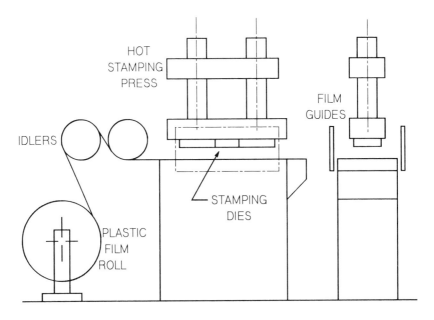

Figure 9.11. Tolerance analysis case study.

- As part of the setup, the plastic film guides are adjusted for specific film widths.

On the basis of the dimensions and tolerances, compute the maximum and minimum limits in print margin from the *left* film guide (see Figure 9.12). Using a statistical analysis, compute the average deviation from the mean. Assume a 6-inch-wide film for this case.

The print margin is defined as the dimension measured from the top edge of the printing to the film edge.

THEORY OF OPERATION (See Figure 9.11) Rolls of plastic film are loaded onto an unloading reel and go through two idlers (or straightening rolls) to the hot stamping press. After the printing operation the press also die-cuts the plastic as required. The operation requires a minimum clearance (0.010 inch) between the film and the guides. Because of this clearance, the plastic film will track between the left and right guides.

MODES OF FAILURE The goal of printing is to minimize the variation in the print margin. One way to accomplish it is to keep the track guides as close as possible. But this can lead to the film jamming between the guides, which is unacceptable.

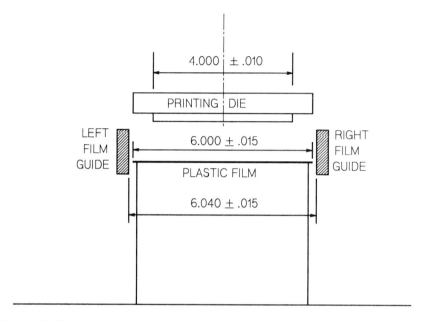

Figure 9.12. Press printing section, nominal dimensions and tolerances (inches).

The next approach is to purchase film with a narrower tolerance band. This will increase the cost and hence the price the customer has to pay. The tooling for dies can be better, but it will also increase the cost.

The last approach is to accept a calculated amount of wastage by rejecting parts produced at the maximum print margin.

SOLUTION To solve this problem, a tolerance analysis will be performed to compute the maximum and minimum limits in print margin, using the extreme-case analysis. We will also apply statistical analysis to compute the average deviation in print margin from the mean for a given set of conditions.

Calculations require the following to be taken into account:

Tracking of the film in between the film guides
Tolerances of
  plastic film
  width of the printing die
  film guide adjustment

(See Figure 9.12 for nominal dimensions and tolerances.)

Table 9.6 lists the different permutations that are obtained by considering the maximum and minimum tolerances of the plastic film, printing die, and film guides. As seen, there are seven different combinations in the extreme conditions and countless numbers of possibilities in between.

TOLERANCE ANALYSIS CASE STUDY

TABLE 9.6. Permutations for Calculating Plastic Film Tolerance

No.	Die width		Plastic film		Film guides	
	Min	Max	Min	Max	Min	Max
1	×	—	×	—	×	—
2	×	—	×	—	—	×
3	×	—	—	×	—	×
4	×	—	—	×	×	—
5	—	×	—	×	—	×
6	—	×	×	—	—	×
7	—	×	—	×	×	—

For example, in combination 3, the die width would be at the minimum, the plastic film width would be at the maximum and the film guides would be adjusted to the maximum possible.

Minimum print margin. (See Figure 9.13 Referencing the left guide.) To compute the minimum print margin, we select combination #6:

Die width is max = 4.010 inches.
Plastic film width is min = 5.985 inches.
Film guides adjustment is max = 6.055 inches.

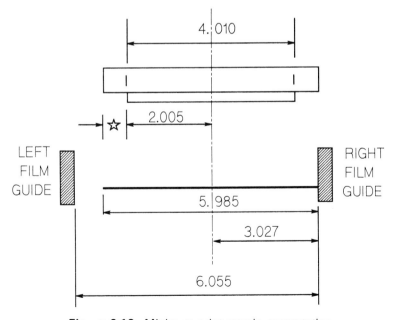

Figure 9.13. Minimum print margin computation.

As stated in the theory of operation, the plastic film will track between the film guides. When it tracks to the extreme right and hits the right film guide, this condition will create the minimum print margin on the left:

Print margin (left) = (5.985 − 3.027 − 2.005) = 0.953 inch

Note that it is a fallacy to assume that the right side print margin is the maximum. Computing the right side print margin with this set of conditions results in (3.027 − 2.005) = 1.002 inch. As will be seen, this is not the absolute maximum print margin.

Maximum print margin (left). (See Figure 9.14.) To compute the maximum print margin we select combination #3:

Die width is min = 3.990 inches.
Plastic film width max = 6.015 inches.
Film guides adjustment is max = 6.055 inches.

In this case we take the condition when the plastic film tracks to the left and hits the left film guide:

Maximum print margin = (3.027 − 1.995) = 1.032 inch

It is important to note in this case that the plastic film width maximum or minimum is not of any consequence. We could equally have selected condition #2.

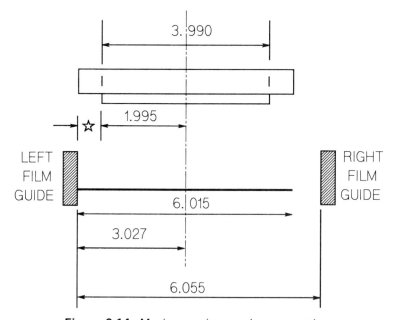

Figure 9.14. Maximum print margin computation.

Average print margin deviation. Using statistical analysis one can determine the average print margin deviation from the nominal dimensions for σ_1, σ_2, and σ_3.

Printed die:	4.000 ± 0.010
Plastic film	6.000 ± 0.015
Plastic film guides:	6.040 ± 0.015

Assuming T (total tolerance) = $\pm 3\sigma$:

$\sigma_1 = 0.010/3 = 0.0033$
$\sigma_2 = 0.015/3 = 0.005$
$\sigma_3 = 0.015/3 = 0.005$

Putting these values into equation (9.1) to compute the average standard deviation yields:

$$\sigma_{avg} = \sqrt{(\sigma_1^2 + \sigma_2^2 + \sigma_3^2)}$$
$$= 0.0078 \text{ inch}$$

The average deviation for $\pm 3\sigma$ will be ± 0.024 inch from the mean, or a total variation of 0.048 inch. This means that 99.7% of the printing will fall within the margin of ± 0.024 inch. For 95.5% of the parts the variation will be two standard deviations, or ± 0.016 inch.

CONCLUSION The variation in print margin from the maximum to minimum is $(1.032 - 0.952) = 0.080$ inch. Using statistical analysis we derive a total variation of 0.048 inch (which covers the 98.8% range). This is considerably less than the maximum print margin. If a study is done of the printing operation and the tolerances of some of the variables can be tightened, then a much narrower distribution can be obtained. This case study also underlines the importance of understanding the process when doing tolerance analysis. If, for example, the printer specifies 0.080 inch variation in print margin to his customers, the probability exists that he may lose jobs because of the high variations in his process (determined by extreme-case analysis). In actuality his variation is much tighter (0.048 inch). This example was used to acquaint the reader with the diversity of applications for this approach and with its simplicity.

9.9 CONCLUSION

In this era of fierce international competition, any new knowledge is too important to be ignored. GDT and tolerance analysis are two tools that offer immediate benefits.

GDT is quickly becoming a universal language for mechanical drawings. ANSI, which has pioneered the work on GDT, has introduced new concepts that obviate the ambiguity that leads to mistakes. It has been aptly said that 90 percent of all problems are communication problems. The new concepts of datums, MMC, LMC, RFS, and controls clearly explain the design intent to the fabricator.

Tolerance analysis has been an indispensable tool since the beginning of the industrial revolution. It is more important now because of the new materials and processes that are coming into being every day. With the reduction in the life of products because of changing technologies, companies are under tremendous pressure to reduce the time to market of their products. Tolerance analysis uncovers any mistakes in design before they become major disasters.

SUGGESTED READING

ANSI Y14.5M-1982. New York: Industrial Press, 1982.

Al Wakil, Sherif. *Processes and Design for Manufacturing.* Englewood Cliffs, N.J.: Prentice-Hall, 1989.

Punccohar, D.E. *Interpretation of Geometric Dimensioning and Tolerancing.* Dearborn, Mich.: Society of Manufacturing Engineers (SME) Press, 1990.

Wade, Oliver R. *Tolerance Control in Design and Manufacturing.* New York: Industrial Press, 1989.

CHAPTER 10

Design For Manufacture of Printed Circuit Boards

Over the course of the last several years, the phrase design for manufacturability (DFM) has joined the growing list of three-letter acronyms that frequently appear in business and manufacturing publications. Among the benefits of DFM that have already been realized by the first converts are included faster product development cycles, reduction in time to market (TTM), and lower fabrication costs.

Because of the increasing pressure to shorten TTM, designers and product engineers no longer have the luxury of using the trial-and-error of prototypes to fine-tune a new design. In addition, the farther down the development cycle a new product is, the greater the difficulty in introducing a design change for any reason. By applying DFM principles during the early stages of the design process, a company can avoid a mismatch between performance requirements of a product and manufacturing capability.

Application of DFM also results in reduced manufacturing cost. At many companies, it has been estimated that 75% of the manufacturing cost of a product is determined by the design. As a result, it may make more sense to optimize the design using DFM principles rather than work on improving manufacturing processes to fit the design.

Figure 10.1 illustrates the effect of process improvements versus product improvements as a means of getting higher performance at optimum cost. The curve shows product manufacturing cost versus design complexity. The expectation is that higher complexity is purchased at higher cost. In order to reduce the cost, manufacturing is often expected to implement additional automation, new process technology, quality assurance, and so on. These have the effect of moving the response curve to the right so that the same design can be manufactured at lower cost or so that higher complexity (and theoretically, higher performance) can be obtained at the same cost.

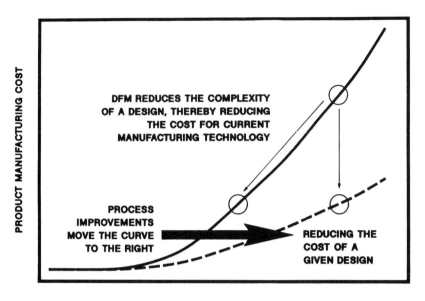

Figure 10.1. DFM and design complexity.

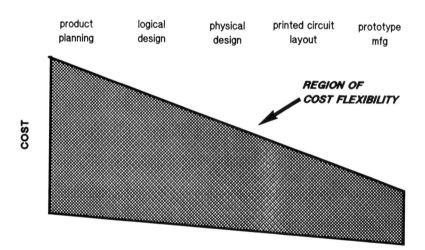

Figure 10.2. Printed circuit board design stages.

DFM has a different effect on this curve. If the complexity of the design can be reduced using DFM without sacrificing performance, cost can be reduced without introducing additional changes in manufacturing. DFM helps product designers make optimum use of the current state of manufacturing technology.

In electronics manufacturing, decisions made during the early phases of product planning and logical design have a greater impact on final cost than is often realized. Because these selections tend to get locked in, there is greater flexibility in reducing cost if DFM is applied at these critical stages (Figure 10.2).

Everyone seems to agree that DFM is something worth doing. Unfortunately, like other popular acronyms such as JIT and SPC, few seem to know what it means or how to make it happen. This chapter summarizes the key elements of a successful DFM program for printed circuit board (PCB) fabrication. While the focus will be on DFM for PCB manufacture, the basic principles and concepts are common to DFM for any product.

10.1 PRINTED CIRCUIT DESIGN

Typically, 10 percent of the total material cost of an electronic product is due to bare printed circuit boards and as much as 90 percent of the material cost is accounted for by printed circuit assemblies (board with integrated circuit components). With the exception of custom logic integrated circuits (ICs), no other electronic system component receives as much attention during electronic product design as does the printed circuit board.

Each stage of the electronic system design process requires decisions that serve to define the final printed circuit design. Product planning and logic assign the partitioning of function between chip-level and board-level and therefore determine the approximate size and number of PCBs. In many cases, a previously defined standard card cage leads to a board length and width requirement. The type of integrated circuit packing used in the printed circuit assembly, whether simpler through-hole dual-in-line package (DIP) or surface-mounted quad flat pack (QFP), is an important determining factor for the PCB design.

Electrical performance characteristics such as timing requirements, controlled impedance, noise control, and maximum current/voltage are also important performance drivers. These inputs lead to decisions regarding dielectric material, metallization type, board thickness, dielectric thickness, construction layup, and hole sizes. Finally, the design reaches a CAD operator who has some flexibility in determining the trace width and layer count during schematic capture, layout, and routing.

Not surprisingly, it is often possible to create more than one PCB design that meets the specified performance requirements. When faced with multiple printed circuit design alternatives, how does one decide which is "best"? Most people would agree that it is the one that results in the lowest board price. No one wants to spend more money for a given level of performance than they have to. Once the minimum performance level has been defined, the goal is to minimize cost.

Another important consideration is time. Because of TTM pressure and the increased circuit density of PCBs, many designers have been forced to sacrifice manufacturing cost in favor of faster design. Such is the case with autorouting, which is the system for connecting the traces of the PCBs during the design stage. The strategy is to set up the CAD system's autorouter using defaults or "standard" design rules and then accept whatever combination of trace width and layer count results.

10.1.1 PCB Design Alternatives Case Study

The following characteristics describe a fairly simple combination of components and schematic from a real product design:

56	gates
50	devices
59	equivalent ICs
182	signal lines
290	non-supply connections
27.56	square inches board area

Depending on the CAD system and the skill and patience of the individual designer, there may be an almost infinite number of possible PCB layouts for this case. Here are three actual design alternatives:

	1	2	3
Layers	6	4	4
Vias	130	270	234
Line length	411.726 in	466.512 in	411.644 in
Lines/spaces	12/12 mils	8/7 mils	6/6 mils
Total holes	791	931	895

Assuming that each of these is equally capable of delivering the necessary performance, this case study will illustrate the best design for the PCB fabrication.

Typically, design selection criteria are a combination of several considerations. If the number of design alternatives is small, it may be possible to obtain a price estimate or quote for the fabricated board. However, as we shall see later, using board price to make this decision can be misleading, especially for a captive PCB supplier. Price is often based on prevailing market conditions or the desire of a particular vendor to attract or discourage business to match their product offerings.

More frequently, designers apply their intuition or experience with similar board designs from earlier projects. This informal database is very difficult to maintain, particularly as experienced designers retire or move on. Therefore, most companies have attempted to set down these principles as "best practices" or design recommendations. These may also take the form of a company-wide printed circuit design standard that must be used for all new products. The Institute of Printed Circuits (IPC) has published several industry-wide standards that are widely used for these purposes.

It is interesting to note that although the people who create the standard may consider it simply as a set of guidelines, the users of these specifications frequently view them as something inviolable handed down on stone tablets.

There are several major drawbacks to these traditional selection criteria. First, they provide a purely subjective basis for comparison of designs. The attitude of "We've always done it this way" has become more difficult to defend in today's competitive business environment. Second, experience and intuition tend to vary considerably from person to person and design to design. The result can be a design standard with more exceptions than rules.

Finally, such criteria often fail to take into account design interactions. It does not require twenty years of experience in printed circuit fabrication to know what makes a board less manufacturable. Designs should have as wide a trace as possible to avoid yield loss due to electrical opens. Designs should have a low layer count to reduce material cost. Every drilled hole is a potential electrical open or short, so hole count should be minimized.

Each of these "best practices" is absolutely true when considered individually. Now, go back to the design alternatives and try to make a decision. It cannot be done unless the interactions and trade-offs between these design features have been characterized.

Although these drawbacks may not seem formidable in the simple example cited above, one can easily imagine more complex situations. The most common decision that a printed circuit designer must face on every project is whether to add signal layers or go to a higher wiring density per layer, using narrower lines and spaces. In other words, when does the higher material and processing cost of more layers offset the lower yield of finer lines? A quantifiable, systematic approach to DFM is required to resolve this dilemma.

10.1.2 Evolution of DFM for PCBs

The concept of DFM as a rigorous, documented program or process is fairly new. Historically, DFM has consisted of a collection of recommendations or best practices based on the collected experience of senior experts. In most cases, no formal system exists. The typical approach is to assign the best people to the job, review the design with manufacturing, prototype it, and hope for the best.

The beginning of a more systematic approach to DFM can be traced to the fundamental research by Geoffrey Boothroyd and Peter Dewhurst of the University of Rhode Island. They published their book *Product Design for Assembly* detailing the methodology they developed. The Boothroyd and Dewhurst system analyzed each part of an assembly from the standpoint of necessity of existence of separate parts and their ease of handling, feeding, and orienting. The results were an estimate of the assembly time and a rating for design efficiency. This was the beginning of quantitative DFM.

Hitachi's Production Engineering Research Laboratory (PERL) developed a technique for examining a mechanical design called the *assemblability evaluation method* (AEM). This system was the result of basic research on the cost of motions, joining methods, and group technology. AEM rates the ease of assembly of a design against a set of standards, to produce a score that is a relative measure of how easily the parts will go together. General Electric obtained a license from Hitachi to use the method, improved it, and developed English-language training materials (1983).

The Hitachi/GE studies led to an assembly model that examines the manufacturing process as a series of individual steps. The probability of introducing a fatal defect at each step is determined by the nature of the process and specific design elements. Penalty points are assigned to processes or elements that introduce difficulty, such as an operation that requires registration and alignment of several pieces at once.

A high final score suggests a low assembly yield. If the score is over an arbitrary threshold, the design may be placed in a "low-buildability" category that may require a review with the product engineer or direct intervention of production and engineering during assembly.

Figure 10.3 shows the process of applying this DFM system to a new product. The initial design is analyzed to identify the necessary assembly steps and a processibility rating is calculated from the sum of penalty points. A review and iteration loop provides on opportunity to redesign and minimize the score. The score is a measure, or metric, that can then be used to quickly rank design alternatives.

PRINTED CIRCUIT DESIGN 241

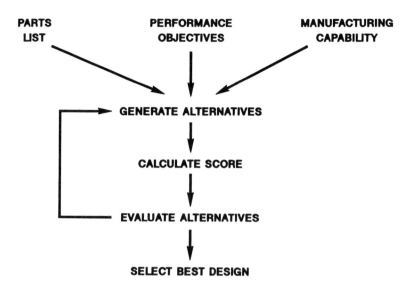

Figure 10.3. Applying DFM to a new product.

Other attempts have been made to extend this idea of assigning penalty points to manufacturing processes or design elements. This has proved to be especially useful for mechanical assembly operations. Unfortunately, these models are inappropriate for nonassembly manufacturing such as printed circuit fabrication for the following reasons.

Summed algorithms are inappropriate for printed circuit fabrication because of the statistical uncertainty associated with the individual process steps. Each step has a probability of inducing a fatal defect that depends on a unique subset of design elements. Furthermore, the likelihood of a single step, such as drilling, causing a defect is extremely dependent on previous steps such as inner-layer imaging or lamination. Individual steps cannot be viewed independently.

Interactions and trade-offs between design elements are difficult to characterize. For example, the diameters of drilled holes should be large to avoid high-aspect-ratio plating and the higher cost of reduced drill stack height. On the other hand, for a fixed pad size, a smaller hole results in a larger hole-to-pad annular ring. This annular ring acts as insurance against drilled hole breakout or reduced conductor width that can result from extreme misregistration. Therefore, the effect of hole size on manufacturability and cost cannot be expressed as a simple linear relationship (see Figure 10.4).

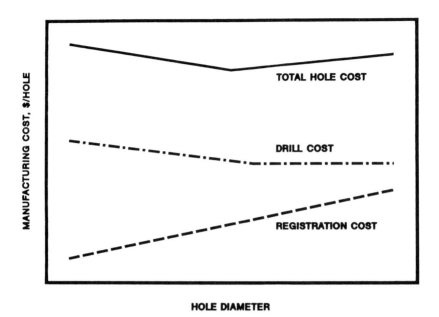

Figure 10.4. Manufacturing cost versus hole diameter.

10.2 DFM PROGRAM REQUIREMENTS

The DFM system for printed circuits should include:

- Information based on actual fabrication capabilities (not intuition).
- A process for comparing design alternatives (prior to layout if possible) using a quantifiable manufacturability or cost metric.
- Methods for determining the effect of design features—individually or in combination—on final PCB cost.
- Methods for determining the optimum design within specified performance requirements.

Furthermore, the program should be guided by the following principles.

- *Information is neutral.* The final selection of the "best" alternative should be left to the designer or product engineer. The problem with design standards is that they dictate right and wrong answers when in fact many of the so-called rules can be broken with a cost penalty. This is important, because the rules are constantly being challenged by demanding product designs.

- *Information is accurate and current.* The printed circuit manufacturer must be willing to back up the DFM system with a consistent pricing strategy. That is, if a designer using the DFM process concludes that option A is more manufacturable than option B, then A had better cost less than B. By the same token, if a particular fabrication process is expensive to operate, this should be reflected in the DFM metrics. Countersunk holes may indeed be difficult to manufacture but, unless there is a cost penalty associated with the process, this information has no place in a DFM system. Since manufacturing capability will improve over time, a support system must be created to periodically validate and update the DFM process.
- *Information is quantifiable.* DFM cannot be subjective. Previous methods based on years of design or manufacturing experience, but unsupported by metrics, are beginning to break down.
- *The DFM process must provide incentives.* The use of DFM is directly related to its ability to clearly identify incentives (see Figure 10.5). Design engineers will not apply DFM unless there is a demonstrated advantage. The most significant incentive is product cost, so the DFM program must be grounded in an accurate pricing model. Credibility is directly related to the DFM/price relationship. A weak relationship will ensure failure. A strong relationship will establish trust and incentive.
- *The DFM process must be easy to learn and use.* Metrics must be unambiguous and relevant and the DFM process must match the current engineering design process. If it takes a few days or even a few hours to get feedback on the manufacturability or cost of a design, it will not be used. The process should be tailored to suit anyone who is involved

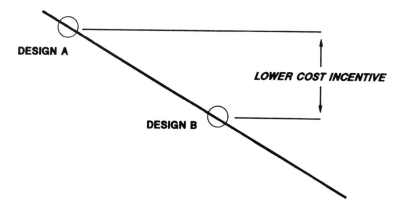

Figure 10.5. Incentives versus design alternatives.

in board design, whether CAD operator or R&D project manager, without requiring a crash course in PCB fabrication.

Two metrics are key to meeting these objectives of the printed circuit DFM program: relative cost and design complexity.

10.2.1 Relative Cost

The most important metric in any DFM program is a measure that provides product engineers with a tool for comparing the manufacturing cost of alternative designs. An obvious choice as a cost metric is actual board price; however, this is not as attractive as it may seem at first glance. Printed circuit suppliers, even captive shops, are extremely reluctant to share pricing algorithms. Furthermore, board price depends on the profit margin, which is usually based on the prevailing market conditions and can therefore be quite volatile. In any case, the objective is to compare designs, not estimate price for a single design.

The DFM cost metric should model those elements of board price that are directly affected by the design. Any fixed cost that is design-independent should not be included in the metric.

The manufacturing cost model for printed circuit fabrication will vary greatly from one supplier to another. However, many operations use the concept of a cost center with a specific rate applied to each unit that passes through the center. For example, if it costs $1 million per year to operate and maintain a cost center and recent history indicates that 10,000 units per year will be processed, then the rate for the center is $100 per unit. For a printed circuit shop, cost centers might include:

- inner-layer processing (imaging and etching)
- inner-layer inspection
- lamination
- drill
- outer-layer processing (plating, etching)
- soldermask
- graphics
- final inspection and electrical test
- other special processes (e.g., gold connector fingers)

These centers contribute to the design-dependent manufacturing cost of a board. A two-sided board will not require inner-layer processing or lamination. Similarly, many shops use inner-layer inspection only for very dense or controlled-impedance designs. The type of soldermask (screened, dry film, or liquid photoimageable) will determine which rate is applied.

DFM PROGRAM REQUIREMENTS

In addition to the process cost, there is material cost for the copper-clad dielectric, prepreg, soldermask, gold plating, and so on. The operating cost of a manufacturing facility will include overhead for such things as administration and marketing. However, these are usually applied evenly across all board products and are therefore not design-dependent.

By examining the list of cost centers, the board features that affect the cost of the design can be identified. The number of layers will determine how many times the inner-layer processing rate is applied and the material cost of copper-clad dielectric and prepreg. Since the drill cost center rate is often expressed per hole, this cost will depend on the stack height, which is in turn based on the board thickness and hole diameter. Most cost centers process full panels (such as the 18 inch × 24 inch size), so the more images or boards that can be designed on a single panel, the smaller is the per-board cost.

The PCB DFM cost model provides a conversion process from board design features to manufacturing cost. In order to keep actual cost in dollars confidential, a relative cost metric can be created to express the same absolute relationships. This metric should be scaled and normalized so that differences in relative cost correspond to relative differences in manufacturing cost as calculated from the cost centers.

To calculate the relative cost of a proposed design, a product engineer must provide some basic characteristics such as the number of boards per panel, number of layers, metallization type, soldermask type, and dielectric material type. A sample worksheet is shown in Figure 10.6. The user is

RELATIVE COST WORKSHEET

1	number of layers		enter Table 1 value
2	metallization type		enter Table 2 value
3	drilling factor		enter Table 3 value
4	dielectric material		enter Table 4 value
5	soldermask type		enter Table 5 value
6	screened graphics?		if yes, enter XX per side
7	additional panel processing		enter sum from Table 6
8	relative cost per panel		add lines 1 through 7
9	boards per panel		
10	base relative cost per board		divide line 8 by line 9
11	additional board processing		enter sum from Table 7
12	unyielded relative cost per board		add lines 10 and 11
13	estimated yield		refer to yield model
14	relative cost		divide line 12 by line 13

Figure 10.6. Relative cost worksheet for PCBs.

prompted to use a look-up table to find the relative cost of each characteristic. The total relative cost is simply the sum of the individual contributions.

The relative cost data in the look-up tables is derived from the cost center rates and the material cost. For example, layer count is converted to relative cost using the following equation:

$$\text{Relative cost} = (\text{total material cost} + \text{number of innerlayer pieces} \times \text{innerlayer rate}) \times \text{conversion factor}$$

Relative cost provides a standard measure of the contribution of individual design features to overall board manufacturing cost. It allows product engineers to compare multiple design alternatives. It measures cost differences in percentages, not dollars.

The combination of a relative cost worksheet and look-up tables is simple to use and does not require an in-depth knowledge of printed circuit processing. It also lends itself quite readily to a software spreadsheet format.

The missing element in this relative cost algorithm is the one that most people would describe as manufacturability. For some designs, yield can be a more significant contributor to overall manufacturing cost than raw materials or overhead burden. Production scheduling must compensate for a low-yield design by increasing the number of panels launched. The added cost of scrap is usually passed to customers as a higher price per board or appears as a yield variance. Without an accurate yield prediction model, the supplier risks overcharging the customer or absorbing the cost of defective boards.

In order to ensure cost recovery, board suppliers must be able to estimate fabrication yield from design information.

10.2.2 Complexity

Printed circuit shop tooling departments routinely provide manufacturability reviews of new board designs prior to production release. These reviews are successful in identifying major errors such as spacing violations or missing features. However, in most cases, the yield has already been determined by decisions made far upstream and it is too late to significantly alter the design. While factors that contribute to yield loss are well known (including high layer count, fine lines, and small holes), higher performance unavoidably requires selection of features that create less-manufacturable boards.

Designers and product R & D engineers need a tool to evaluate the effect of feature selection on yield at the early stages of the design process, thereby minimizing the board cost for a set of performance requirements. A particularly effective way of providing this tool is to express the manufacturability of a design technology set with a single standard metric. The metric is defined by a mathematical expression containing values of the significant design elements. The yield prediction model then becomes simply a functional relationship between fabrication yield and the manufacturability metric. This method allows several different design alternatives to be compared quantitatively to determine the yield (or cost) improvements associated with selected design changes.

An added advantage of this metric is that it provides a standard measure of a design or finished board for general use. The metric offers a common basis for discussion between designers, marketing, finance, and manufacturing and can be used to follow increases in board technology over time.

A fabrication yield model should provide a reasonably accurate estimate of scrap rate, be easy to use, account for interactions between design elements, and provide some predictive capability for new designs.

During the past several years, there have been a number of attempts to create a yield model using board design elements to provide the best fit to historical data. By applying regression analysis and a linear model, combinations of variables can be correlated to yield. Excellent fits have been obtained; however, these models are extremely cumbersome and lack the expressive power of a simple manufacturability metric. Although quite accurate, it is unlikely that such a system would be commonly used. Furthermore, because the models were empirically derived from yield of current part numbers, there is no guarantee that the results can be extrapolated for future designs.

A possible design complexity factor is based on five variables: board area, layer count, hole count, minimum feature spacing, and tolerance. When the yield of a selected group of designs was plotted against the logarithm of the complexity factor, the data appeared to follow a nonlinear curve described by an equation known as the Weibel function. This function is commonly used in product reliability studies to represent failure mechanisms (see Figure 10.7, courtesy of Happy Holden, Hewlett Packard Corporation).

This combination of a relatively simple metric and a functional relationship between the metric and fabrication yield appeared promising, but a more accurate model was required. The strategy chosen was to examine the printed circuit fabrication process to determine the sensitivity of yield to design elements and improve the definition of complexity.

The development of a complexity metric should be guided by an empirical study of the influence of design elements on fabrication yield. First, each printed circuit manufacturing process should be examined to uncover possible sources of yield loss. Then, the most common fatal defects observed in manufacturing are investigated to determine probable design and process-related causes.

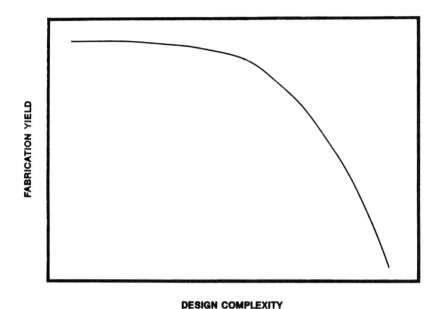

Figure 10.7. Fabrication yield versus design complexity.

The brief overview of printed circuit fabrication processes and common defects focuses attention on design features that have a probable impact on yield. Thus, design elements are identified for a statistical analysis of variance of production yield data for current part numbers.

The overall fabrication yield of a printed circuit board is limited by the maximum capabilities and normal process variations of the individual steps. The key processes which introduce yield loss are:

Image transfer
Copper etching
Lamination
Drilling
Metallization
Soldermask

Figure 10.8 summarizes some of the basic limitations of these fabrication steps and the associated critical design features that account for the occurrence of defects.

An analysis of PCBs scrapped over an extended period revealed that the majority of fatal defects fall into three categories: electrical opens, electrical

DFM PROGRAM REQUIREMENTS

FABRICATION PROCESS	YIELD LOSS ISSUES	CRITICAL DESIGN ELEMENTS
image transfer	imaging resolution artwork registration particle contamination	trace width/spacing line length board area
copper etching	over-etching	trace width line length
lamination	material shift layer registration interfacial delamination	layer count board area innerlayer thickness
drilling	drill wander mechanical damage lack of through hole connectivity	hole count board thickness hole diameter & annular ring
metallization	smear removal lack of through hole connectivity	board thickness hole diameter
soldermask	misregistration	soldermask clearance

Figure 10.8. Limitations of fabrication steps.

shorts, and soldermask defects (e.g., cracking, flaking, or loss of adhesion). The following is a list of possible design-related causes of these defects.

Defect	*Critical design feature*
Electrical open	Trace width
	Line length
	Board area
	Layer count
	Hole count
	Board thickness
	Hole diameter
Electrical short	Spacing
	Line length
	Board area
	Layer count
	Hole count
Soldermask	Soldermask clearance

In order to verify the apparent effect of these features on yield and obtain an estimate of their relative significance, tooling and work order yield data should be collected for production part numbers. Although a production shop does not provide a controlled environment, owing to process variability, there is considerable value in using historical data on actual boards rather than an experimental test vehicle that may not resemble a functional design.

Yields of PCBs are typically assumed to follow a normal (or Gaussian) distribution function with a mean value and symmetric distribution of individual observations on either side of the mean. If all work orders were the same size, the mean would be used as the "average" yield of the design. More commonly, the average yield is determined by a volume-weighting procedure that simply divides the total quantity of defect-free boards by the total lot start quantity over a selected interval of time.

The use of the Gaussian distribution to fit work order yields is based on the assumption that, for a given board technology set and fabrication process, there is a "most likely" or nominal yield. The range of actual yields may be described by the standard deviation or "three-sigma" limit about this mean.

The tendency for observations to follow the Gaussian distribution (the central limit theorem) is well established for deviations from a nominal value. For large populations of boards, the physical dimensions and electrical properties (e.g., board thickness, finished hole size, line width, or impedance) will be normally distributed. Each step in the fabrication process introduces an error and the sum of these errors is the deviation of the measured value from the nominal value.

However, production yields should not be expected to behave in the same manner. Yield is essentially a binary state measurement. A single board will either pass or fail on the basis of acceptance criteria. For example, a board can be scrapped because of extreme soldermask misregistration leading to mask on a pad. The minimum soldermask clearance will follow a normal distribution, but if the clearance exceeds a threshold value (zero), that board fails. Other fatal defects caused by catastrophic events such as a short due to board contamination cannot be represented by random process error.

Figure 10.9 shows a series of histograms of production yields for typical printed circuit boards. In each case, the solid line is the best fit to a Gaussian distribution to the yield data. The poor correlation reveals the fact that the histograms are skewed with a long tail at low yields and no values greater than 100%. This shape suggests an alternate statistical representation given by the gamma distribution function. The mean and standard deviation of the gamma distribution have the same meaning as the more familiar Gaussian values. The function maximum corresponds to the value with the highest frequency of occurrence (the statistical *mode*).

Figure 10.10 displays the histograms as Figure 10.9 with the best fit of the gamma distribution indicated by a solid line. In the analysis of design elements used to develop the complexity metric, work order yields are reduced to their gamma mean values.

DFM PROGRAM REQUIREMENTS

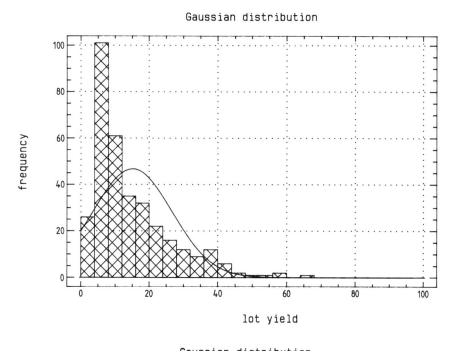

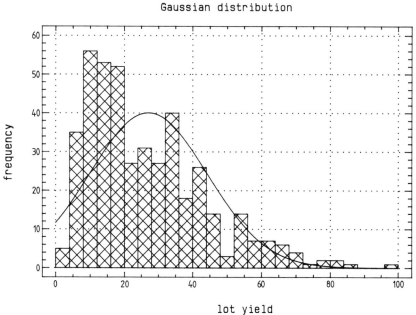

Figure 10.9. Histograms of production yields for typical printed circuit boards.

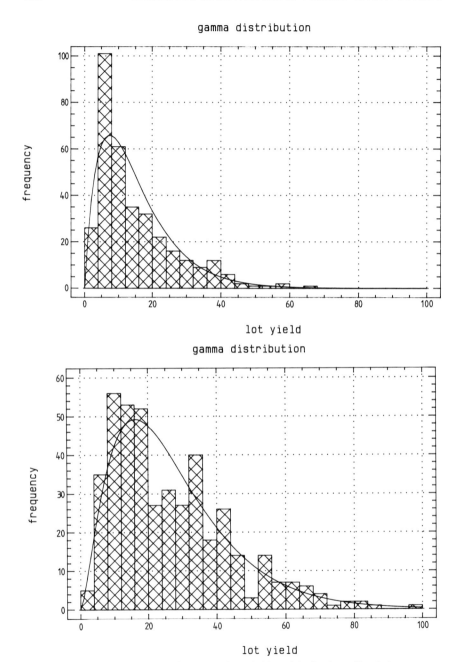

Figure 10.10. Histograms of production yields with the best fit of the gamma distribution indicated by a solid line.

PERFORMANCE MEASURES

A complexity metric and yield prediction model will be revealed from a statistical analysis of production yields and design elements. The accuracy of the model should be good; however, the overall predictive capability is not the issue. The goal is to provide some measure of yield of a design, not to predict the yield of any given lot of boards. The yield estimate will be used as part of the relative cost calculation that compares board designs.

Recently, IPC has begun an effort to define DFM metrics for the printed circuit industry. Complexity indicators are currently being defined for board design, board fabrication, board test, board assembly, board assembly test, and assembly repair. These metrics, or figures of merit, rank a design from easy to impossible on a scale of 1 to 100. One of the objectives of this program is to create a profile that describes a board design or a printed circuit supplier. By comparing the design profile with a database of supplier profiles, a match can be identified that will ensure optimum manufacturability.

10.3 PERFORMANCE MEASURES

The one issue that may seem to have been forgotten in the previous discussion about DFM is that the PCB design must perform a function. Many product engineers will refuse to discuss DFM because they are convinced that their design is the only one that can do the job. It is vital to the success of the DFM program that the performance drivers of the design are understood so that alternative designs can not only be rated, but suggested as well.

Printed circuit board performance can be described in terms of two general drivers: connectivity and electrical performance. The most common expression of connectivity is inches of wiring per square inch of circuit board (in/in^2). While it may be defined in various ways, it is always calculated from board features such as track (or number of traces between grid points) and layer count. Although widely quoted, this metric is actually not very useful in printed circuit design or DFM. Inches per square inch does not determine the wiring demand of a component library and circuit schematic. Rather, it is a measure of the wiring capacity of the combination of trace width, spacing, signal layer count, and pad size (or technology set). These are only known after layout and routing. Hence, inches per square inch cannot be used to describe the necessary connectivity required and thereby guide these critical decisions.

Printed circuit designers need a tool to predict what combinations of track and layer count are likely to successfully route a given component library and schematic prior to layout. Most designers currently make an estimate of track and signal layer count from prior experience. The CAD software autorouter is set up with "standard" feature dimensions (usually from a design specification) and allowed to work away. After a reasonable interval, the layout is checked for remaining disconnects. If a significant number of signals are still unconnected, the usual procedure is to add another pair of signal layers. From a DFM perspective, however, it may make more sense to increase track by reducing trace width, spacing and/or pad size. Without the means to describe the required performance and estimate the necessary board technology, there is no basis on which to do a DFM cost analysis until after layout.

A connectivity performance model should use the information available to the designer after schematic capture. A number of theoretial studies (primarily by Donald Seraphim at IBM) have demonstrated that wiring demand can be calculated from the number of I/Os per component, the number of components, and the approximate spacing between components. All of this should be known prior to layout. If the capability of the CAD system and its operator can be expressed as an efficiency factor, then wiring capacity and board technology sets can be estimated.

How does this work? In one sample algorithm, the designer calculates the total number of equivalent ICs (EICs) from the component library and a look-up table. By dividing the EICs by the board area available for routing, the EIC density is determined. The EIC density serves as the connectivity metric. Another table displays the possible board technology sets that correspond to the EIC density value, perhaps ranked by relative cost or probability of successful routing. This table is created from a survey of the historical capabilities of designers and CAD systems.

Another major requirement for printed circuits is electrical performance. The copper traces on the board are an important part of the entire circuit, particularly for high-speed applications. This is especially true for designs requiring controlled impedance, where signals must be terminated in the characteristic impedance of the device in order to avoid reflections and false triggering.

For a given performance requirement such as 50 ohms $\pm 10\%$, there are a number of combinations of trace width, dielectric thickness, and model type (e.g., stripline, microstrip) that can provide the necessary impedance. Again, the design engineer can benefit greatly from a simple table that displays some possible combinations ranked by relative cost. Similar tables or graphs could be created to guide other design decisions for electrical performance, such as crosstalk, RFI/EMI protection, rise time degradation, capacitance, and resistance.

10.4 OVERALL PROCESS

How are all these metrics and models used? The printed circuit DFM process is summarized in Figure 10.11. The performance requirements for the design

CONCLUSION

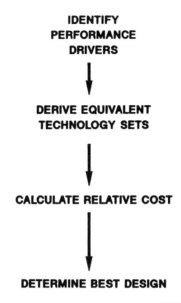

Figure 10.11. DFM program for PCBs.

are defined using driver metrics such as connectivity and characteristic impedance. Next, equivalent technology sets representing combinations of design features are identified. Finally, the optimum technology set is selected based on manufacturing yield and relative cost.

Let us go back to the design example cited. Using the complexity/yield prediction model and the relative cost metric, the three alternatives can be scored and compared:

	1	2	3
Log complexity	8.7	9.3	9.4
Estimated yield (%)	91	86	85
Relative cost	17	15	15

In this case, the savings in material cost realized by reducing layer count from 6 to 4 outweighs the slightly lower fabrication yield due to narrower line widths. If relative cost is scaled and normalized against actual dollars, options 2 or 3 represent a 12% cost reduction over option 1.

10.5 CONCLUSION

How does a DFM program get started? The tendency observed in several infant DFM programs is to try to take on the whole problem and become intimidated by the daunting task. The fact is that experienced product designers generally know what they are doing. They only have difficulty in a few key areas. The first step in developing a DFM program for a product is not to make a list of manufacturing likes and dislikes. The first step is to examine the current design process to understand the most common decisions that must be made to create the design. Where are the decision points? When do design options begin to appear? What information must be provided to assist in making the decision?

From the standpoint of the manufacturer, there are two types of DFM "don'ts":

- Things that cannot be done because the manufacturing process is physically incapable of delivering the requested performance. For example, the plating process may be incapable of depositing 1 mil of copper in a through hole with aspect ratio of 10:1. That is, not a single board would have the desired plating thickness.
- Things that could be done but at low yield or high cost (i.e. not an engineering impossibility). Say the designer wanted a 5:1 aspect ratio. The plating process might be able to deliver this performance, but at a cost of lower throughput due to lower current density. In addition, a sampling/inspection scheme might need to be implemented to insure that shipped boards will meet the 1 mil thickness specification.

It is important to help the design engineer understand the difference between the two. Unfortunately, most design rules or standards fail to do this.

From the standpoint of the designer, there are two types of design rules:

- Rules that would rarely be broken because the product engineer would have no reason to break them. Obeying these rules will not compromise board performance and they are broken only because of ignorance of error.
- Rules that will always be pushed to the limit for performance reasons.

The DFM program should always focus on the second type of design rules. In fact, it may be appropriate to eliminate design standards altogether and replace them with a system of cost-impact analysis. The designer would be free to select any combination of possible features and calculate the overall manufacturability of the design using the relative cost metric.

Developing a DFM program means understanding the current process costs, developing a fabrication yield metric (if appropriate), and combining it with a relative cost metric that helps designers make the low-cost decision, but does not replace the price quoting scheme. It is important that the design alternative with the lowest relative cost is also the one with the lowest price. If possible, it is desirable to understand the performance of the product well enough to offer alternatives with the same performance, but ranked according to relative cost.

The PCB DFM program should be backed up with a support plan. Unless the DFM information is constantly updated, the program will die. There must be inputs from all functions:

R & D	New products and services
Manufacturing	New processes or yield improvement affecting cost or yield metrics
Finance	New cost information
Marketing	Competitive price information

The ultimate goal of any DFM program should be to integrate with design for assembly, reliability, servicability, repairability, testability, and so on. A combined "-ilities" or sometimes called "DFX" program gives the product engineer a means to choose optimum designs that have taken into account the often conflicting demands of system-level performance.

To summarize, the success factors for any DFM program are:

Listen to designers.
Tailor the content accordingly.
Base the system on metrics.
Be consistent with cost/price strategy.
Recognize design performance requirements.
Provide a training program.
Provide on-going support and verification.
Work towards ultimate integration with DFX.

SUGGESTED READING

Band, G. "How to design, procure and manufacture quality printed circuit boards at the lowest cost." Proceedings of NEPCON WEST, Des Plaines, Ill., 1990, pp. 433–458.

Boothroyd, G. and Dewhurst, P. *Product Design for Assembly*. Wakefield, R.I., 1987.

Duck, T. "Concurrent engineering for PCB product development." *Surface Mount Technology*, Best of Surface Mount '90, pp. 13–18.

Fiskel, J. and Hayes-Roth, F. "A requirements manager for concurrent engineering in printed circuit board design and production." *Second National Symposium on Concurrent Engineering*, February 1990.

Hwiszczak, R. and Johnson, R. "Achieving producibility in PWB design using CAD/CAM tools." Proceedings of NEPCON WEST, Des Plaines, Ill., 1990, pp. 1249–1262.

Jodoin, C. "Simultaneous engineering in PCB design for manufacture." *PCB EXPO '90*, January 1990.

CHAPTER 11

Reliability Enhancement Measures for Design and Manufacturing

New product design for reliability has undergone significant changes in the recent past. The use of computer-based analysis for electrical and mechanical designs has increased, especially with the wide use of computer-aided design (CAD). Once the geometry of the part or its electrical connectivity and components are captured in the CAD system, computers can perform the analysis function automatically, using selected software analysis packages. Previously, the mechanical or electrical elements had to be described in tabular input form to the software, making the operation tedious and error prone.

The increased use of these analysis packages has become widespread with the use of leased or purchased software, and through networked engineering workstations. For electrical designs, the use of SPICE modeling for transistors, operational amplifiers, and other analog circuit designs has proven to be very beneficial in simulating circuit parametric response to inputs such as gain, delay time, and output signals. For digital designs, several simulators are available that can (1) validate the Boolean logic functionality of the design, (2) simulate design properties such as delay time or possible electric signal race conditions, and (3) verify the functional testing of the design. For mechanical designs, the use of computer analysis packages such as finite-element, moldflow, thermal, and vibration analyses has enhanced the validation of the design before commitment to actual testing of the design prototypes, which reduces the number of prototype iterations, and therefore decreases the total development time and cost of the project.

These analysis methods impact the traditional methods of achieving high reliability through exhaustive testing of the design prototypes. This chapter will examine the current methods of enhancing reliability in the light of computer design analysis, and will explore traditional and new methods used by world-class manufacturing companies to increase new product reliability.

11.1 PRODUCT RELIABILITY SYSTEMS

These reliability systems have historically been based on the performance of electronic components. Figure 11.1 is known as the *bathtub curve* of component failure rates versus time. The bathtub shape indicates three different phases of reliability during the product life cycle. There is an initial "infant mortality" period where components that are marginal because of defects in their original manufacturing cycle or use in subsequent manufacturing operations will become defective. This phase is followed by the "useful life" period, where, once infant mortality failures have been completed, the electronic product will exhibit a level failure rate, with randomly occurring failures based on the stress of the normal operation of the product. The third phase, called the "wearout" phase, occurs when parts in the product begin to reach the end of their useful life as outlined by the product specifications and fail according to the levels speculated by the design team. The bathtub shape comes from the congruence of the three factors: production and process failures, labeled "quality" failures in Figure 11.1, which cause infant mortality, are hyperbolic; stress-related or random failures occur at a level rate throughout the life of the product in the "useful life" stage; wearout failures

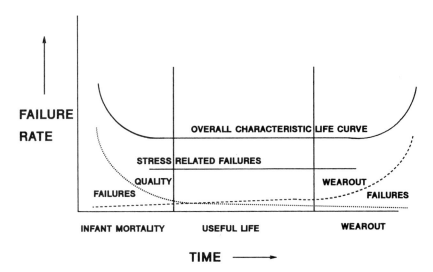

Figure 11.1. Bathtub curve of component failure rates.

are determined in the design phase to last according to component and product specifications.

In actuality, not all electronic components and parts follow the bathtub curve pattern exactly. The type of component will determine which of the three causes of failures will predominate. Wearout is especially important in mechanical or electrical components with parameters that change with increased use. An example is print heads that mechanically wear because of the friction with paper. Another example is the cathode ray tube that has a specified useful life from its manufacturer because of phosphor depletion. In both cases, the manufacturers will specify a certain lifetime use, and the components should be replaced when appropriate.

Passive electronic components such as resistors, capacitors, and inductors tend to follow the stress-related failure curve. Their failures do not have an infant mortality pattern, but they occur randomly. This may be due to random variations in the process of their encapsulation during manufacture or in the assembly process into printed circuit boards (PCBs), where they might be exposed to random variations in component insertion, placement, or soldering.

Active electronic components such as integrated circuits (ICs), both linear and digital, have a much higher failure rate than passive components. ICs tend to have higher failures as device complexity increases. Complexity could be measured in terms of the manufacturing processes used for fabricating the ICs, such as line widths and density of transistors in ICs. Some of these devices do exhibit the early failure or "infant mortality" syndrome. The causes are due more to processing problems in the fabrication of the IC's devices. This is a complex process involving many steps of etching, cleaning, deposition and welding, and use of harsh chemicals and solvents at elevated temperatures. A misstep or a random deviation in the fabrication process timing or control could cause some of the ICs to fail prematurely, leading to the observed early failure effects.

11.1.1 Bathtub Curve and Burn-in

Belief in the utility of the bathtub curve of component failures has prompted many companies to take steps during the manufacturing cycle to attempt to capture the infant mortality period in-house. Electronic products are subjected to a burn-in period before they are shipped. During burn-in, products are turned on, either on the shelf or in specially designed environments or high temperature, and sometimes with cycled temperature and power. This philosophy is in conflict with modern manufacturing philosophies such as just-in-time (JIT), which stresses the fast turnaround of orders.

There are many disagreements between the manufacturing and quality departments of companies on the effectiveness of the burn-in. The quality

PRODUCT RELIABILITY SYSTEMS

department will point to continuing failures during burn-in as the proof that it is needed, while the manufacturing department will experiment with shipping products with and without burn-in and collect field failure data that will show that burn-in does not statistically influence the field failure rate. The argument is made more confusing since the clever use of statistics can bolster either of these conflicting arguments. Another counter-argument to burn-in is that the stress of applying temperature and power cycling in the burn-in process might shorten the useful life of the product and increase the onset of wearout.

Most electronic products show a declining failure rate in the field versus time. The decline is fairly linear during the life of the product, as in Figure 11.2. There is a significant reduction of failures during the later part of the product life cycle in the field, with some product showing as much as 50% decrease in the rate of failures after one year. It is very difficult to achieve a significant reduction of failures through extended factory burn-in. Even if burn-in is performed at elevated or cycling power and temperature, this "accelerated life" environment will only increase real-life simulated period by a factor of 2 or 3. At this rate, 4–6 months of burn-in would be required to reduce failure rate by 50%.

However, world-class manufacturing companies usually perform some burn-in of their products, but its duration and environment varies. The objectives of burn-in are similar for many companies:

- Burn-in is good for creating real failures in-house for the engineers to analyze in controlled conditions. Even in the most advanced quality-conscious companies, precise information about failure mechanisms and the availability of the failed components is lacking. As an example, it is a common experience that 50% of the PCBs of electronic products

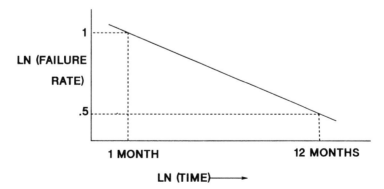

Figure 11.2. Failure rate versus time.

returned to the factory because of field failures are "No Trouble Found" (NTF)!
- Burn-in is a good method of preventing a future catastrophic component or process failure that otherwise might go undetected. Because manufacturing processes are getting shorter, the probability of discovering a time-dependent failure, also known as a "time bomb", is lessened.
- Burn-in can provide the feedback data necessary for supporting a total quality control (TQC) program to improve design, process, and incoming-materials problems. The defects found in burn-in can be analyzed, grouped into a Pareto analysis, and targeted for enhancement.

For many companies, the amount of final product burn-in time could be reduced by burning-in the electronic assemblies or individual components before they are used in the product. If burn-in is applied earlier in the manufacturing cycle, the cost of repair or replacement of the defects will be lower. An effective burn-in program should be tailored to the individual components and products, while a universal burn-in program is likely to be ineffective at best, and could result in damage to the product.

11.1.2 Product Reliability Strategy

When designing new products, the reliability targets and projections are an important part of the product design strategy. The reliability objectives of new products will be influenced by the following factors.

The cost distribution of the enterprise. The reliability objectives could be achieved by balancing the different alternatives of the design versus life cycle costs. These choices could be increased development costs in the implementation of design analysis programs and methods of robust designs, versus additional life cycle costs. These costs may be due to incoming inspection; tests such as in-process, burn-in, and final manufacturing; expensive and higher specification components; and warranty and field repair. It has been shown in many companies that additional investment in designing for reliability in the development phase is much smaller than that in increasing reliability through production test and repair.

Customer expectations. Customer expectations of the reliability of electronic products are increasing significantly as more companies are able to significantly increase their products' reliability over time. These expectations include the cost of ownership of the product, which is the expected failure rate, usually measured as mean time between failures (MTBF), and the expected repair time, expressed as mean time to repair (MTTR). The warranty support period has become an important factor in the competition for customers in the electronics industry. Lifetime or extended warranty is offered for many products, which, in the absence of a good reliability strategy, could be very detrimental to the long-term financial performance of the company.

User environment. Customer use of the product is important in setting and achieving reliability targets. Specifying particular temperature, humidity, and also electrical power surge limits for product operation eases the design function by shifting the burden to the customer. Such is the case with products designed for eventual use on a factory floor, which have to be modified for electrical noise, hazards, and shock generated by manufacturing equipment. Static electricity protection is very important for electronic products used in office environments: Even antistatic carpets can generate a significant amount of electricity, causing electronic component failures.

Expected service life of the product. For electronic products, the total life cycle is decreasing because of technology advantages that offer more productivity savings to the customer. Ten years is a common benchmark of expected service life for products, including field support after the product has been obsoleted. Supporting older products is becoming more expensive as parts are discontinued more quickly and their availability for older product support becomes difficult. In addition, there is a lack of adequate test equipment and technicians knowledgeable in older technologies and products. The service strategy could be formulated to encourage early retirement of older products by exchange programs and increasing support costs for older products.

Liability of the company in terms of the expected performance of the product. Specific industries such as financial services and aerospace applications have a large negative impact or poor reliability in terms of contractual agreements of MTBF, MTTR, and uptime. Lawsuits stemming from medical electronics products with poor reliability can be a serious factor in continuing operations in that market.

These elements should be considered when formulating the reliability strategy of new products. This strategy should outline the following steps during the development phase: product reliability objectives; the component selection process; design analysis methods; reliability design review policy; stress tests to be performed and analysis of the results. In the manufacturing and field support phases the manufacturing and warranty targets and objectives should be outlined, including incoming inspection, in-process and final test, burn-in, service and warranty analysis.

11.1.3 Design for Reliability

Designing reliability for new products is an evolutionary process. Aggressive reliability targets must be set and a process developed to meet them. There should be continuous quality improvement through learning and emulating proven designs to enhance new product reliability. In many companies, electronic products will increase their reliability over the product life cycle by as much as a factor of 5 to 10 times. When the company is finally producing a very reliable product, it is obsoleted and replaced with a new generation, which too often starts at the original reliability level of the old product. A design for reliability process should emphasize the following items:

- Set reliability measures and goals consistently and make reliability objectives the highest priority for the design team. Reliability measures such as MTBF, MTTR, and warranty costs should be clearly defined at the earliest part of the design project. Knowledge of the existing products' reliability levels should be available in order to estimate the reliability design task. Reduction of manufacturing variability, design complexity, and product environmental, electrical, and mechanical stress should have a positive impact on reliability.
- Select reliable components, processes, and technologies. The company's internal record of components and processes is the best guide in the selection of new products. If new components and processes are to be used, they should be qualified with exhaustive stress tests to failure. Analysis should be performed on each failure to determine the cause and, if appropriate, to modify the design to eliminate such failures.
- Minimize part count, both for reducing assembly time and for increasing reliability. Every part is a potential reliability failure, and the fewer parts there are, especially complex ones, the less chance there is for future failures.
- Analyze new designs for marginal and stress life tests at each major phase or checkpoint of the product design life cycle. Analyze every failure throughout the design cycle, by collecting all failed parts and recording the conditions and parameters of each failure. Keep a record of failed parts, perform failure analysis to find cause, and design them out.
- Use the design review process to hold reliability reviews to eliminate poor design practices and to review progress on meeting reliability goals.

11.2 DESIGN TOOLS AND TECHNIQUES FOR ENHANCING RELIABILITY

These design tools and techniques have been developed by world class manufacturing companies through many product generations and reliability enhancement programs. They are based on product reliability studies and methods for enhancing reliability estimates. The tools can be used by the design team to determine the reliability strategy for new products.

DESIGN TOOLS AND TECHNIQUES FOR ENHANCING RELIABILITY

One of the important metrics for new products is the reliability estimate, as expressed by mean time between failures (MTBF). This metric is based on the historical evidence of component failures, which are the most common source of electronic product failures in the field. Of all common sources of product failures, component failures could account for more than 80%, followed by workmanship problems of fabrication and assembly such as those caused by bad solder joints and cabling problems.

Component-based MTBF is calculated by taking the electronic component list of the product and the quantities used and multiplying them by the failure rate of each type of component. The failure rate is based on 10^6 hours of operation, and can be obtained from standard tables such as the MIL Handbook 217E or American Telephone and Telegraph Corporation (AT&T) estimates of component failures. The sum of the multiplications forms the total product failure (Tf) per 10^6 hours. MTBF is calculated by taking the inverse of the total failures. The probability of successful operation for 100 hours without failures [$R(100)$] is determined by approximating a Poisson distribution with a mean of 100/MTBF.

Total failures per 10^6 hours:

$$Tf = \sum (\text{component type} \times \text{quantity used} \times \text{component failure per } 10^6 \text{ hours})$$

Mean time between failures:

$$\text{MTBF(hours)} = 10^6 / Tf$$

Probability of successful operation for 100 hours without failures:

$$R(100) = e^{-100/\text{MTBF}}$$

This method is a good starting point for determining reliability estimates. However, these estimates are based on historical failures of components, and do not reflect recent improvements in component fabrication process quality. In addition, these numbers could be distorted by the company's incoming inspection and supplier certification strategy, as well as by the proper use of the components in the new products. If the total number of products in the field were large enough, the company could develop the component failure rate table from historical field performance of its own products.

11.2.1 Design Review for Reliability

The design review process is normally conducted by the individual project engineer and is attended by team and outside experts as well as management personnel. It serves as a formal checkpoint to ensure that the design meets its intended specifications and functionality.

An important part of the design review is the proper application of component types used for the specific design. Component suppliers issue individual specifications as well as typical application examples, which should be compared to the design being reviewed. Any similar use of components in past products should also be investigated and performance analyzed from factory and field data such as proper electrical input, output, and power loading of integrated circuits.

Another element of the design review process is the determination of the design margin between the component parameter specifications and the design requirements. Selecting the proper margin is a balance between the higher cost of a wider-specified component versus improved reliability. The techniques of robust design of experiments could be used to determine the optimum choice of component design margins.

Design reviews should also include an analysis of the potential failure modes and ensure that the design is fail-safe: The failure of a controlled on/off heating circuit should cause it to shut down, thereby preventing potential fire damage.

Potential problem analysis can be performed using one or both of two methods: fault tree analysis (FTA) and failure mode and effect analysis (FMEA). Both methods are similar, with FTA using a top-down method of determining the general ways in which the product could fail and graphically showing the possible causes of failures down to the component level, while FMEA uses a bottom-up approach of tabulating all components and their potential defects.

FMEA provides a formal mechanism to resolve the potential problems by (1) identifying all possible ways a component can fail, (2) ranking these ways with a risk priority number (RPN), (3) suggesting a corrective action, and (4) calculating the resulting improvement in the RPN. The FMEA methodology begins with identifying each component and listing the potential failure modes, the potential causes and effects of each failure, the initial control conditions, and conditions contributing to the RPN. These conditions include the probability that the failure takes place (occurrence), the damage resulting from the failure (severity), and the probability of detecting the failure in-house (detection). This analysis suggests a recommended action for eliminating the failure condition by assigning a responsible person or department and recalculating the RPN.

FTA is a relational diagram showing possible sources of defects and their interrelationships. It uses logic-type symbols of AND and OR gates to show intermediate and final defect symptoms. Examples of FMEA and DTA are shown in Figures 11.3 and 11.4.

DESIGN TOOLS AND TECHNIQUES FOR ENHANCING RELIABILITY 267

**FAILURE MODE AND EFFECTS ANALYSIS (FMEA)
FOR DESIGN AND DEVELOPMENT**

PRODUCT_____	MODEL AFFECTED_____	PAGE___OF___
PRIMARY PRODUCT RESPONSIBILITY_____	SCHEDULED START-UP DATE_____	FMEA DATE (ORIGINAL)___(REV)___
OUTSIDE SUPPLIERS AFFECTED_____	PROJECT ENGINEER_____	APPROVED BY_____

					CURRENT STATUS						RESULTING STATUS				
PART NAME PART #	PROCESS FUNCTION	POTENTIAL FAILURE MODE	POTENTIAL EFFECT(S) FAILURES	ITEMS	POTENTIAL CAUSE(S) FAILURE	CURRENT CONTROLS	OCCURRENCE	SEVERITY	DETECTION	RISK PRIORITY # (RPN)	RECOM ACTION & STATUS	ACTION TAKEN	OCCURRENCE	SEVERITY	DETECTION	RISK PRIORITY # (RPN)	PERS RESP

Figure 11.3. Failure mode and effect analysis (FMEA).

11.2.2 Stress Analysis for Reliability

The techniques of analyzing stress points for reliability are focused on identifying the "weak points" in the design that will be the most probable to fail. These techniques are different from the software-based analyses of designs that validate the design versus original specifications and functionality.

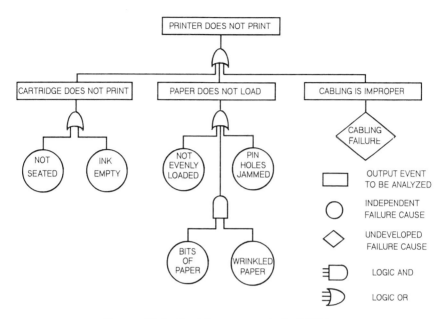

Figure 11.4. Decision tree analysis (DTA).

Thermal mapping for electronic circuits and stress imaging for mechanical designs are similar, and require specialized equipment. They produce graphical presentations of future weak points by highlighting areas of above-average heat or stress. These areas could be potential failure points for electronic or mechanical designs and, once identified, can be eliminated by redesigning the product to redistribute heat or reduce mechanical stress by altering the geometry or material of the product. Component performance and lifetime will decrease with increasing temperature and elevated mechanical stress, and the use of heat sinks and electric fans can be beneficial in lowering the ambient temperature of the product.

Another strategy for reducing potential weak points is the derating of components and parts used in the design. An example is the use of an 1/4 watt resistor in an electronic circuit, instead of 1/8 watt. This derating strategy could potentially lead to a significant product cost increase, which could offset the benefits of increased reliability.

Another analysis to be performed in the design stage is the mechanical review. An important example is the tolerance stackup analysis (see Chapter 9). If tolerance analysis is not performed, problems in mechanical parts fitting together in production will be evident even if prototypes appeared to have been made to fit together and to be mechanically and functionally correct. This is evident because when the product is made in production volumes, manufacturing variability will produce much wider tolerances.

Chemical and environmental stress reviews have to be conducted separately to ensure proper selection of components, technologies, and manufacturing processes that can withstand the highly corrosive temperature and humidity variations in certain industries such as the chemical and oil industries.

11.3 PRODUCT TESTING FOR ENHANCING RELIABILITY IN DESIGN AND MANUFACTURING

The product performance changes in conditions of external stress. Testing in extended temperature, humidity, power cycling, shock, and vibration will also yield important data in terms of product performance degradation and failed components. The products and components can be analyzed to determine cause and designed out. These tests could be performed in the product design cycle in order to characterize product performance and to reduce manufacturing test.

Stress testing can be focused on particular components, technologies, and manufacturing processes. The stress level can be increased by cycling the stress factor over time. The conditions of these stress tests can be obtained from the Department of Defense and the military organizations (DOD and MIL specifications) and as well as several standards organizations in the United States such as the American National Standards Institute (ANSI), the Electronic Industries Association (EIA), the Institute for Interconnecting and Packaging Electronic Circuits (IPC), and the International Electrotechnical Commission (IEC).

Stress testing can be focused on component types as shown in Table 11.1.

11.3.1 New Component Qualification Process

Design teams should use existing and proven components and processes as much as possible. New components and processes present an unknown reliability factor in new product designs. Implementing a qualification strategy for new components, equipment, and manufacturing processes is important in maintaining a high reliability level. In addition, if the component is being supplied from a new vendor, a supplier qualification process should also be implemented. The steps for implementing a component qualification program are as follows:

- Determine all specifications: Although specifications are supplied with new components, some important parameters are defined as typical,

TABLE 11.1. Stress Testing

Stress type	Failure mechanism	Component type
Voltage	Dielectric breakdown Hot electron injection	Capacitors, insulators MOS devices
Current	Metal fusing and migration	Switches, relays and contacts ICs & power semiconductors
Temperature	Chemical reaction Ion contamination	Plastics, electrolytics, Linear and field-effect devices
Temperature cycling	Differential expansion Thermomechanical failure Condensation (with humidity)	Solder joints, especially SMT Plastic package devices
Moisture	Ionic contamination Corrosion and dendrites Leakage	Plating and sealed assemblies Plastic package devices PCB-bound analog circuits
Shock and vibration	Loosening and misalignment Creepage and embrittlement Conducting particles	Mechanical assemblies Adhesives and welds Electrolytic capacitors

minimum, or maximum. Examples are timing characteristics of digital ICs, which, if not properly characterized, can work properly in prototypes but cause race conditions in the electronic circuits when used in production quantities and variabilities. These characteristics have to be qualified by running a large number of samples to characterize the parameter.

- Analyze the construction and performance of the samples supplied. A thorough analysis of the methods of fabrication and assembly for components and mechanical parts should be undertaken. either independently by the design team and its support organization (component engineering), or by using independent analysis laboratories. A good knowledge of the component fabrication and assembly processes (such as IC and PCB fabrication) should be acquired by attending courses and seminars on the subject. A small analysis laboratory could be set up, with magnification and polishing equipment to expose IC and PCB inner construction, in order to section off components and observe their manufacturing methods and modes of failure.
- Stress test the new component samples to failure, to learn of the failure mechanism, using the stress chart supplied in Table 11.1. Determine the worst-case failure mechanism, and redesign to eliminate it.
- Learn the potential uses of the components in the product and determine the stress and failure mechanisms expected. Choose components and processes that are the least sensitive to expected stress in the intended product use.

PRODUCT TESTING FOR ENHANCING RELIABILITY   271

- Keep all failed components and document all failure mechanisms. Make suppliers aware of the failures and negotiate mutually accepted quality and performance levels. Use quality tools such as the process capability index (C_p, see Chapter 4) to develop a partnership with the supplier.
- If the component is being supplied from a new vendor, use a supplier qualification process to evaluate the supplier's quality control procedures. Request process control documentation and make supplier plant visits to ensure proper operation of the supplier's processes.

11.3.2 Product Qualification and Validation Testing

Every new product should be subjected to stress testing in order to isolate weak designs and components, and to have them redesigned out. Product qualification and validation can follow procedures established by the company's internal guidelines, or those suggested by government or industry associations. These tests fall into three categories.

Margin tests apply varying stresses to the product in terms of temperature, pressure, humidity, altitude, shock, and vibration. The stress could be cycled, with varying transition slopes between the cycle maximum and minimum.

Life tests should be performed on new products to determine useful life at highest allowable stress as specified. The life tests should be simulated under several accelerating conditions such as highly elevated temperatures. This test is sometimes referred to as *test to failure*. Component failures should be corrected and the life test continued until a catastrophic failure occurs, at which time the test is aborted. This test requires extensive product samples of more than 20 units and more than 1000 hours duration.

Field tests should be performed under actual field conditions, especially if the environmental factors are not controlled.

All failures during these tests should be documented, and failed components analyzed for the cause of failures. There should be a depository of failed components. A list of failures and analysis reports should be tabulated by the design team. The status of each component is to be kept up to date as to the disposition of failure, suggested solutions, and a note of the responsible engineer to resolve the problem.

11.3.3 Manufacturing Process and Reliability

Reliability estimates are made for new products on the basis of a reasonable quality control of the manufacturing materials and processes. However, many quality systems measure the status of the new product's attributes and parameters, leaving other manufacturing process-related problems undetected. These problems are known in the reliability jargon as *workmanship defects*.

Process procedures are normally not as well documented as product assembly and test procedures, and the reliance is on the goodwill of the operators and the memory of supervisors to maintain process integrity and enforce good manufacturing practices.

Many of these manufacturing process-related sources of poor reliability could be avoided by the use of common sense as well as good and effective training of the workforce. The normal test cycle for electronic products will not be able to uncover some of the process problems that can result in potential field failures. Performing exhaustive testing to isolate these problems is impractical. Though automation can solve the majority of process defects, it is very expensive and not suitable for all industries. Prevention by process documentation and the training of the workers and their managers is the most effective method of improving workmanship failures in the field.

Some of these process-induced potential failures involve poor adherence to process specifications, such as not adjusting the production equipment to the proper setting or shortening the time for production operations. Poor work habits such as contamination by food in assembly, poor handling of work parts, and illicit repair procedures can also cause failures. Improperly designed or adjusted equipment can cause partial damage to the product. The product design team should be well aware of the manufacturing process, and can either design their products more robustly or suggest more documentation and control of the manufacturing process.

The categories of these process-induced reliability failures are divided into the following.

Electrical failures are the most common. Electrostatic discharge (ESD) protection is very important, especially with the use of MOS devices in the product. Some static discharge has been shown to cause partial damage to the intermetallic layers of ICs, causing the ICs to pass the factory tests but fail prematurely in the field. ESD protection should be provided with antistatic storage and transportation containers, work mats, grounding of equipment and operators, and by requiring operators not to wear nylon clothing. In addition, proper environmental control of temperature and humidity is critical. There should be periodic surveys of the workplace for potential static sources and their elimination.

Other electrically generated failures include improper test equipment design or maintenance that causes generation of "spikes," which are excessive current or voltage noise. Other failures could be caused by improper sequencing of test voltages, continuity tests on active circuits, dangling high voltage test probes, and improper grounding of chassis and subassemblies.

Mechanical failures usually originate from cracking, bending or abrasion of components, parts, and assemblies. These can be induced by poorly adjusted equipment such as connector riveting, auto insertion, and placement of electronic components, when either pressure or alignment of workheads is incorrect. Critical manual assembly operations should be provided with measured tools to properly set the rate and pressure of work such as torque wrenches, adjustable pressure glue dispensers, and adjustable current welding equipment.

Shock and vibration can be damaging to assembly and joining processes such as welding, soldering, and gluing. Excessive shock and vibration could be generated from poorly designed handling and transportation systems, such as conveyor lines, transportation carts, and automatic inventory storage and retrieval systems. Other process systems such as spray cleaning, painting lines, or ultrasonic cleaning systems (especially for PCB boards) could add potential damage to electronic products.

Contamination of product materials and chemicals, either through continuous use or brought in by contaminated parts or operators, can be especially damaging to electronic circuits and material finishes. These include fluxes, solvents, encapsulating compounds, plating and cleaning solutions, and paints and sprays. They should be monitored through incoming inspection either by chemical or physical analysis or by actual use on workpieces. In-process monitoring should be performed using analysis and indirect measurement such as specific gravity, ionic conduction, and pH meters.

Environmental related failures causing damage to electronic processes and components, such as improper temperature and humidity variations, should be prevented by proper controls. These could reduce the shelf-life of materials and components, and could cause premature oxidation of component leads, rendering them difficult to solder and making for poor grounding of metal parts.

Extreme heat and cold used in the production or repair of electronic circuits is undesirable. Proper setting of solder tips, "hot air" guns, and surface mount technology assembly repair stations is very important to prevent damage to components. In addition, spray coolant used in electronic troubleshooting should be avoided. These conditions cause thermal shock to dielectric components and could eventually make them fail prematurely.

The design team should be aware of these process-induced reliability failures, and include a process review as part of the design for reliability.

11.4 DEFECT TRACKING IN THE FIELD

Defect tracking from field failures is important to determine the performance of current products and to design defects out in future products. Collecting defect data requires a substantial investment in a defect reporting and tracking system.

Defect tracking depends on the warranty and exchange systems used by the company. Warranty data is a good source for keeping track of defects, as the customer will report all problems during the warranty periods. After the warranty period is completed, some of the products will continue to be serviced by the company through maintenance contracts, while others will be serviced by the customer and therefore their defects will not be reported back to the tracking system. In that case, the customer might use the company's exchange program and this could become another source of reported defects.

Several problems will adversely affect the defect reporting and tracking system: Integrating data from the various sources of field offices and exchange programs might be difficult. All repair and exchange program facilities should operate on a uniform set of defect and repair codes. These codes can then be reported back to the defect tracking system in the factory and tabulated by defect type, assembly number, part number, and product serial number. A small number of defect types will dominate, but redesign of assemblies and parts to eliminate them might not be economically justified.

The knowledge gained in the defect tracking system should be available to new product design teams in order to promote use of the most reliable parts and assemblies and to avoid high-defect ones. A common IBM new product release philosophy is that "A new product has to exceed or equal the quality and reliability of the product it replaces."

11.5 SUMMARY

Designing new products for reliability is an evolving process. The use of design analysis tools has decreased the reliance on exhaustive testing to determine the quality and reliability levels of new products. The use of stress testing and testing to failure of new components, processes and products is important in order to find defects, analyze why they occurred, and design them out.

Most companies use some burn-in for their products in production. Although it has been shown that there is no real infant mortality period for new products, and the burn-in process does not capture early failures, most world-class manufacturing companies are using burn-in as a source of collecting defects for the internal total quality programs, thus increasing their confidence in their products and processes.

The most important elements of improving reliability are the following: (1) using proven design, techniques, and processes; (2) evaluating designs for reliability by the proper use of components and analysis of the design; and (3) collecting all failures and failed components, determining the causes of defects, and designing them out.

SUGGESTED READING

Engelmaier, W. "Is present day accelerated cycling adequate for surface mount reliability evaluation?" IPC-TP-653. Proceedings of the IPC, Lincolnwood, Ill., September 1986.
Haviland, R. *Engineering Reliability and Long Life Design.* Princeton, N.J.: Van Nostrand, 1964.
Ireson, W.G. *Handbook of Reliability Engineering and Management.* New York: McGraw-Hill, 1989.
Kao, J. "A new life measure of electron tubes." *IRE Transactions. Reliability Quality Control.* PGROC-7, April 1957.
Mann, R. et al. *Methods for Statistical Analysis of Reliability and Life Data.* New York: Wiley, 1974.
Omdahl, T. *Reliability, Availability and Maintainability (RAM) Dictionary.* Milwaukee: ASQC Quality Press.
Wiebull, W. "A statistical function of wide applicability." *Journal of Applied Mechanics*, Vol. 18, 1951.

CHAPTER 12

Tools for DFM
THE ROLE OF INFORMATION TECHNOLOGY IN DFM

Conceptually, a DFM program consists of three components: people and their organization; design and analysis processes; and information technologies that support these organization and design processes. Historically, U.S. businesses have focused their attention on moving material (i.e., parts) smoothly and rapidly through manufacturing to customers. Now, time to market is driven by the ability to develop a product; DFM is concerned with moving a design into manufacturing effectively and efficiently. Design information is the "material" of the design-to-manufacturing transfer. This chapter will focus on the role of information technologies in effective coordination, use, and transfer of design information, providing a catalyst in facilitating and sustaining DFM.

This chapter has two goals. The first is to describe information technology's role in DFM. DFM and information technology concepts will be discussed; information technology requirements will be identified. Next, specific information technologies that are critical to sustaining a consistent DFM practice will be covered.

The second goal is to provide a framework for planning and implementing information systems to support DFM. These concepts and the DFM team concepts found in the earlier chapters will be used to develop guidelines for planning and implementing DFM in an electronics manufacturing company. Lessons learned from other DFM projects and building on the experience to define next steps will be covered.

12.1 INFORMATION TECHNOLOGY'S ROLE IN DFM

12.1.1 Market Pressures Drive the Need for DFM Information

In a free enterprise, customer purchases drive companies; companies either respond to customers and make what customers want, or they fail and go out of business. R. E. Gomory writing in the *Harvard Business Review* states:

> The U.S. is learning now the hard lesson taught the rest of the world earlier this century: Product leadership can be built without scientific leadership if companies excel at [the] design [process] and the management of production.
>
> (Gomory, 1989)

Gomory says that the first things high-technology companies must do to be competitive are: (1) design for manufacture, and (2) reduce their time to market; in short "pulling the right know-how into the development of products at the right time." This represents a shift from previous management thinking, which focused on reducing direct labor and cutting material costs. Because of this shift, information technology's role in the electronics business is changing.

12.1.2 Information Technology's Changing Role

In the broadest sense, information technology is a tool for providing information (not just collected raw data) for better decision making at all levels, and a vehicle for improving people's and organizations' roles in the business. Tom Peters (1987, p. 611) summarizes information technology's role this way:

1. It [an information system] provides critical information that the firm sees the worker as a...problem solver.
2. The widespread availability of information is the only basis for effective day-to-day problem solving...
3. Sharing information on the front line inhibits the upper-level game playing.
4. Visible posting of information radically speeds problem solving and action taking.
5. Information sharing stirs the competitive juices.
6. [Useful] information begets more [useful] information.

In the past, information systems have typically been used to collect information for the purpose of preparing periodic reports on the costs and (hopefully) profits of the business. This data was typically gathered and processed in cost-accounting terms for the purpose of tracking the financial health of the enterprise. But that role is changing. According to Hayes,

Wheelwright, and Clark (1988) "Information transfers serve not only to coordinate conversion steps and material flows [their classical role], but also provide the feedback necessary to make improvements in the factory's procedures, process technology, and operating characteristics." Information technology has become essential for what Tom Peters calls "horizontal management: front-line communications across horizontal barriers" (Peters, 1987, p. 612).

These information transfers, or front-line communications, occur between all the participants in the DFM team: marketing, engineering, procurement, production, manufacturing engineering, quality assurance, materials management, and plant management. The information system that provides the means of front-line communications needs to be tailored to work within each company's unique organization and culture.

Today, by contrast, most feedback is in the form of opinions expressed from each department's point of view. Consider this exchange between engineering and manufacturing:

Manufacturing: "You guys can't design it this way; it's impossible to build!"

Engineering: "You guys can't build anything we give you anyway! All you want us to do is design with yesterday's technology so your job is easier!"

Meanwhile, management reviews the company's declining profit margins, and wonders aloud: "Why can't we cut our overhead costs? We just bought eight new CAD stations last year!"

Further, according to Tom Peters, Robert Hayes, Steven Wheelwright, Kim Clark, and others, information has become the catalyst for the learning experience. This learning process is crucial to getting the right knowledge into the product design cycle at the right time (Gomory, 1989). Unlike traditional communications (i.e., "Here it is, now you deal with it"), these communications span multiple functional departments (engineering, manufacturing, marketing, etc.). For example, analytical manufacturing results from previous designs, instructions, and actions is passed to engineering for development of guidelines and evaluation tools to improve the manufacturability of future designs. If each functional area is ever going to improve its performance, it is critical that it have access to the information that reflects and governs the results of that department's efforts. That information, readily available, and quantitatively expressed (to reduce the departmental rivalry through objective measurements) provides a positive catalyst for change. There are several issues that must be addressed here in order to get the significant benefit of any investment in information technology.

While this chapter focuses on the technological aspects of DFM, the discussion will incorporate the organizational structure and design and manufacturing processes in the implementation of DFM systems. It is important not to fall into the common trap of discussing information system implementation only from the technological point of view, such as which networking protocol is the fastest. This has been a common problem, and has resulted in several good ideas failing because their implementation did not consider the users' needs for wide access to common functions, low response time, and high usability. Rather, one should view the discussion of DFM from the point of view of which technological tools or service will provide the quickest, easiest to use, and widest access to enabling information in the practice of DFM. Organizational aspects and business practices will need to be considered in their interaction with DFM technology.

In addition, one needs to recognize that replacing a clipboard-and-pencil analysis with a computer-based system cannot be successfully measured by whether it is faster to type or write, but rather on the value that the information provides to the organization once it is collected and analyzed (i.e., part failure rates provide design engineers and purchasing better ability to assess vendors' products, providing far more benefit than the cost of the data collection and analysis). The greatest value to the business is not in the automation itself, but in the consistency of the engineering-to-manufacturing process and the real-time communication and evaluation by the DFM team. This is enabled by easy, rapid information-sharing among the team members, making DFM a sustainable practice, not a one-shot experiment. This is the learning process in the industrial environment—the key ingredient in successful companies in the 1990s.

Note that most companies implementing DFM measure their progress with cycle metrics that measure improvement from project to project. In contrast, information feedback to users must be near-instantaneous if engineers and manufacturing personnel are going to use DFM methods and achieve significant reduction in time to market and costs of development.

The evolution of DFM starts with understanding of the manufacturing process and its effect on design choices (i.e., evolving DFM guidelines through the use of statistical quality analysis, design of experiments, and simulation tools to understand the design-through-manufacturing process). The evolution of the DFM practice progresses through improving the manufacturing process through simplification (i.e., using cellular manufacturing and statistical process control techniques). In this phase, statistical process control techniques are used to bring the manufacturing process under control. The DFM guidelines will continue to evolve during this phase, since each time the manufacturing process changes, the guidelines will need to be revised. Once a company has a widespread understanding of the factors that govern the manufacturing process, and has the manufacturing process under control, it can then focus on implementing process concurrency (i.e., concurrent engineering) and process integration (i.e., CIM).

Obviously, with this new role, information technology is not a one-time investment or a static system of information. It is a system that allows for wide dissemination of evaluation methods and information, in a format that promotes decision making in time to avoid cost or schedule impacts. In this way, information technology acts as a catalyst for change, encouraging innovation and learning, not stifling it.

12.1.3 Information Flows That Support DFM

A first step in DFM is understanding how design and manufacturing currently work together, how information is passed today, what additional information is required, and where it must flow to and from. This step is more than a postmortem of the passing of the results of manufacturing's efforts back to engineering. To be effective, these results will be analyzed and guidelines will be developed in time for engineering to use them on the current designs in process. Engineering has never really had a consistent mechanism for understanding the impact of its designs nor has it had the quantitative analysis metrics or tools to improve the producibility of designs. Previously, this information flow of analysis usually occurred only during crises, and was seldom either quantitative or properly documented for future use. Instead, a senior manufacturing person reviewed or modified designs to make them producible. U.S. businesses no longer have those 20-year veterans of the shop floor with their manufacturing and engineering knowledge; consistent information transfers must take their place.

An example of an electronics company's functions and the information flows to engineering that are critical to DFM is shown in Figure 12.1. Note that all cost, quality, and schedule data as well as process performance data associated with the part is sent to the DFM knowledge base. For example, component failures are collected by component part number from the shop floor by barcode or ATE (automatic test equipment) and passed by network via the cell control and/or shop floor control system to the quality analysis system. This failure data is aggregated with the rest of its vendor lot and passed over the network into the DFM knowledge base, where it is indexed by vendor, part number, and process. Statistical analysis will provide manufacturing, engineering, and procurement with consistent, quantitative measures of their product development. This reduces the departmental arguments and provides a vehicle for correcting the current design and evaluating future designs for improvement. This same knowledge base is used by everyone on the DFM, or new product introduction (NPI) team in the development process. This knowledge base should not be confused with a parts database; it is far more robust than that. The knowledge base consists of all the parts data mentioned, and includes manufacturing process tolerance

INFORMATION TECHNOLOGY'S ROLE IN DFM

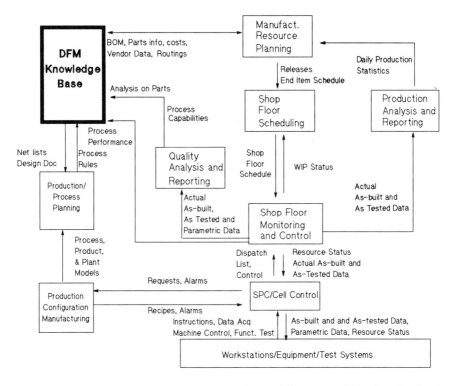

Figure 12.1. Manufacturing information flows. (*Courtesy of Hewlett-Packard Co., Palo Alto, Calif.*)

data as well as product and project design, market, and support goals. It is not necessary for the knowledge based to be one large data base; in fact, in general, information flow is easier and less costly if the knowledge base is an index of where all drawings, designs, analysis, and parameters that relate to the life cycle of a product and process are stored. This approach reduces the level of investment required and the technical complexity when compared with a single large data-storage facility. A single point of storage for all data is similar to storing all the paper that a company uses in the company president's office; no matter how benevolent he may be, storing it there inhibits people's use of the paper and communication throughout the company decreases as a result. A similar result occurs when design or manufacturing develops a design rules or manufacturing guidelines book. The book is a good first step in collecting the information but, because of its format, it does not evolve as easily as an electronic version. As a result, it can quickly lose credibility as a source of accurate guidelines. Furthermore,

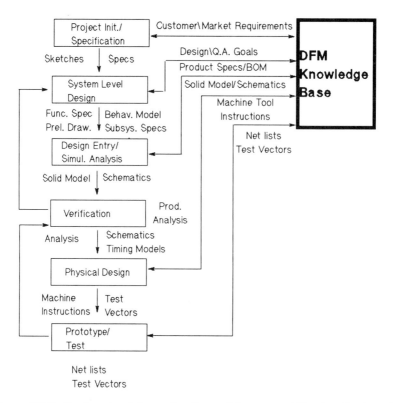

Figure 12.2. Engineering information flows. (*Courtesy of Hewlett-Packard Co., Palo Alto, Calif.*)

a printed document cannot be used as a tool to evaluate a design. Personnel can only refer to it for benchmarks; all evaluation and searching for design data is still left to the designer to do manually. Thus, the paper knowledge base has serious shortcomings and very quickly becomes another inhibitor to communication and learning. The electronic knowledge base provides an index of the current design data, and provides a vehicle for evaluating a design before release to manufacturing.

In Figure 12.2, engineering's information flows to and from the DFM knowledge base are shown. Continuing the example already used, when an engineer needed to select a part, in the past he or she would have searched vendor catalogs for parts, looked at previous designs, or asked a senior designer. This led to the proliferation of redundant parts and a wide variation in design results. With the new knowledge base the engineer enters the design parameters for the part needed (e.g., a 25-ohm resistor with a failure rate less than 100 ppm) and the knowledge base will list all qualified parts, with their failure rates. The engineer can then choose the part that best meets the design goals. Furthermore, the knowledge base is the recipient of all designs and instructions for the shop floor, providing both a conduit and a distribution mechanism to ensure timely information flow. Sophisticated knowledge bases also include evaluation tools to allow analysis of the producibility of a design. Again, note that the knowledge base is shared by all departments to ensure ready availability of common information for design, manufacturing, and procurement decision making.

Fundamentally, all product and process performance goals, design, actual work, and results achieved are recorded and arranged to allow quantitative analysis and use of product, process, and development cycle data by members of the DFM team. What makes this information infrastructure more than simply a raw data shuffling system is the existence of market, product, and process performance goals and metrics against which design and manufacturing results are measured.

12.2 INFORMATION TECHNOLOGY REQUIREMENTS FOR DFM

As an understanding of the company's information requirements is developed, technology requirements will emerge. In the new role of information technology, several critical attributes emerge that DFM information systems must provide:

- *Flexibility* to rapidly change the information flow paths and the information as new requirements emerge. This requirement to change will evolve as Peters' points 4 and 6 (listed earlier) occur through teaming and information sharing.
- *Fast response to users.* Human nature is more comfortable doing familiar tasks that have a known result. Any new method must provide quick response to users to prove its worth over "tried and true" methods. Response times on the screen need to be within tenths of a second and analysis programs need to execute within seconds to minutes. When they do, the results are immediate and the benefit is enormous (Schroeder and Cross, 1990).
- *Consistently easy usability.* Engineers, marketers, and manufacturing will need to use a wider variety of applications, all written by different vendors. The more consistent the human interface, the easier it will be for users to get the benefit of the technology investment. Users seldom spend all their time on one application; therefore, those applications

should be avoided whose human interfaces are complex enough to require frequent usage just to get break-even results (e.g., command driven). The human interface should be simple enough to allow infrequent usage and still allow users to accomplish their tasks.
- *Wide access by all team members.* Many companies have been concerned about the cost of allowing extensive access by their personnel. There has also been concern about the security of proprietary information. Today, lack of information about what customers want and poor data on manufacturing parameters that govern costs are the issues that are primarily responsible for design failure. As for security, while studies have shown that most data loss is inadvertent, any information system or application should provide at least password security and not allow the user to bypass the application, but should not cause the user to wade through multiple levels or obtuse security systems. This will inhibit use and resulting failure of the investment to achieve its payback.
- *Cost-effectiveness.* Almost everyone agrees that technology has some benefit, but how much and when are constant topics of debate. The main issue here is not whether there is a benefit, but how to measure the costs and benefits. The fundamental question every business asks is: "Where, how, and when will the money we spend today be paid back and which investment will provide the best return?" Despite recent research on changing business, information, and investment justification, today the United States still lags behind Europe or Japan in factory investments (Pennar, 1988, p. 102). Peter Drucker, writing on the emerging theory of manufacturing, notes that traditional accounting "can hardly justify a product improvement, let alone an innovation" (Drucker, 1990). He further notes that the primary benefits resulting from the investment in technology lie in the reduction of nonproductive time.

Traditional cost accounting has focused on direct labor reduction. Since direct labor is now commonly less than 6% of the cost of the product in the electronics industry, it is no longer justifiable to focus attention on reducing it. In any event, focusing on direct labor reduction can result in focusing on automating current functions as they are performed today, without simplifying them. Further, with technological change increasing, cost justification of an investment this year should recognize that the task will probably be 10–50% more complex in the next 2–3 years. For example, it is well recognized that available computing price/performance is increasing by at least 20% per year. If one estimates that the design complexity is linearly related and therefore is also increasing at 20% per year, it is clear that significant productivity investments will be necessary, just to keep pace with the complexity! To complicate matters, along with changes in organization and design-to-production flow, this technology investment is predicated on allocation of current resources in the expectation of some future return.

Recent research (Berliner and Brinson, 1987) has led to a whole new model for capital investment. Successful companies will focus investments on tools to help them solve problems more quickly than their competitors and, by capturing good design decisions, developing design rules to speed the design cycle by reducing repetitive decisions and leveraging previous experiences. This becomes a process of continuous improvement instead of the usual do-and-report-done that is the traditional model for business practices and methods. These new tools allow employees to accomplish tasks that the team was unable to do before as well as accomplish the old original tasks with less cost and effort and higher quality of completion. With this new approach to information technology, it is as if one replaces the family car with a helicopter, according to Hayes and Jaikumar (1988). The family is capable of a wider range of activities, not just traveling faster.

The challenge then is to develop realistic estimates of the savings expected and ensure that the proposed investment fits the strategic goals of the company. Investment justification guidelines are outlined later in this chapter.

- *Availability.* In causing a shift in behavior, individuals need to build trust in the use of the new tools. Thus, reliability of the information given is more than the MTBF (mean time between failures) of the hardware. It is the availability of the system when the user needs it, consistent, sufficient performance of the system to provide timely answers, documentation, and field support. Some measures that look at these items are the MTTR (mean time to repair: the average amount of time it takes to get the system repaired) and MLDT (mean logistic delay time: the average time it takes to get the required repair parts to the site). While most vendors do not specify these values, one should ask questions that will provide approximate data about these measurements, such as:

    What is the expected workload or duty cycle for the systems?
    Where is the local technical support office?
    What is its staffing?
    How are support requests handled?
    What level of troubleshooting is expected by the user's company?
    If a part is not in stock, or the problem cannot be solved, how will the problem be handled?

    Since time is critical, these systems need to provide the highest availability for the price paid. Since most problems occur because of external causes (e.g., power failures), buying redundant processors is not a cost-effective answer.

In summary, the attributes necessary for information technology to support DFM must ensure the success of the implementation of the DFM tools through wide access to common functions, low response time, and easy usability. The use of tools (e.g., simulation), and critical information (e.g., historical failure rates and tolerance limits) is promoted by automatic availability. "Automatic" means that every time a user runs the applications he or she is presented with that critical information needed to make a good design decision.

It should be clear that the wealth of functions and information to support DFM could easily take up the DFM team's total effort. Some firms have devoted entire product and process development teams for the task of developing DFM information systems, but this approach has separated the users of the technology from the designers of the technology and has, in general, not delivered a satisfactory result. For example, at a conference of 300 senior managers from Fortune 1000 companies, "managers [were] frustrated by the lack of communication with their information systems (IS) departments, and their lack of involvement in deploying IS resources..." (Freund, 1990). Further, this approach mimics the growth of MIS departments as they exist today—a step backward toward functional specification.

As a result, the DFM team should focus attention on the goals of the project to determine the technology requirements, not the technology of developing the tools. In order to do that, the team will need to choose tools that require the least development, training, customization, and implementation. As with any development project, opportunity cost must always be balanced against development risk. The good news is that, in the case of DFM applications, most of the tools exist today from various vendors. The team can avoid risking the opportunity cost of a product arriving in the market late by investing in off-the-shelf tools instead of building their own. The DFM team will need tools that will enable them to spend their time on achieving the product goals, not building information infrastructures or preventing the access to those infrastructures due to a vendor's proprietary product(s). For this reason, build-it-yourself tool sets generally are not cost-effective. Research has shown that development and support of in-house applications is 4–8 times more expensive than commercially available off-the-shelf products. These observations apply to developing custom solutions in an engineering environment as well as any other environment. Moreover, this tool development effort detracts capital and manpower from development of the product.

Provide the DFM team with the tools they will need to succeed, without inhibiting the product generation process. Fortunately, unlike most systems

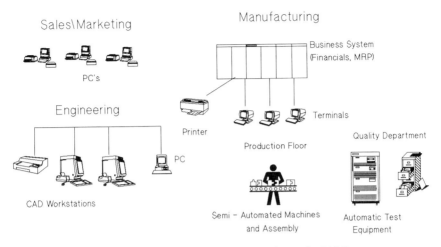

Figure 12.3. Information system infrastructure of a typical U.S. company.

today that are departmental and stand-alone, these DFM tools provide the team with understanding of the manufacturing environment and form the backbone for successful company-wide DFM implementation.

For this discussion, a hypothetical, typical electronics company with islands of automation will be used as a model. It has a manufacturing requirement planning (MRP) and financial system, a few CAD stations, a few personal computers, some semiautomatic manufacturing processes, and a few automatic test systems, all in stand-alone configurations or small stand-alone networks. In a typical company that has been making products for several years, there are already organizational structures, business practices, and information systems in place. The business is typically functionally organized around departments (marketing, sales, engineering, manufacturing, etc.) and each department is measured, judged, and budgeted on its own merits, not those of the products or product strategy that the department supports.

Figure 12.3 shows a typical company's information systems infrastructure. Note that the systems reflect the departmental organization; none are tied together, yet the company must communicate huge amounts of information on a daily basis to produce a product. The significance of these systems is that they provide a starting point; they indicate where process understanding and data may already be stored. It is likely that these systems will pose technological barriers to communication and, in some cases, symbolize organizational isolation. While a discussion of the technical challenge this presents may be interesting, it is the information, not the information system technology, that should be focused on. Management commitment to make the DFM program a reality is the only true universal tool to overcome the existing technological barriers already mentioned. Team building provides a means of developing consensus during implementation, but management commitment will always resolve these barriers. The one set of technologies to prevent the recurrence of these barriers are industry-standards-based technologies. With product as well as information technology changing at increasing rates, this approach represents the safest investment approach. This approach is safe because it allows rapid technology access without the risk of large sunk-cost investments with no growth or recovery. These technologies will be discussed later.

In the evolution of DFM there are many steps. There are a number of projects that will yield early and large paybacks. In the evolution of DFM, the first step is to understand one's manufacturing process. Additionally, since this is an initial step into DFM, building communication and breaking down departmental barriers is a major part of the first steps. Because so much of U.S. business is just getting started in practicing DFM, the rest of these recommendations will focus on technology investments that substantially help begin the process of understanding and improving the engineering-to-manufacturing communication flow. There is a significant amount of learning and organizational adjustment to be accomplished before attempting the advanced stages of DFM: process concurrency and process integration.

One can use information technology very early in a DFM program to promote communication by setting up systems that speed development through easy communication. Voice mail and electronic mail (known as e-mail) networks are excellent tools for enabling communication. Within the scope of this book, a network is defined as a group of hardware and software devices that enable access to applications and sharing of information and resources among multiple users and applications. Most networks consist of the network cable; connections (such as bridges and terminal connectors); communications software on the networked computers; the users' personal computers, workstations, minicomputers, and mainframes; and shared peripherals (e.g., printers). Sophisticated networks will also include network management software and hardware to administer, monitor, and assist in repair of the network. Electronic mail enables the team to communicate across the network with written messages and graphics, enormously speeding up the interoffice memo pace. Voice mail is a means of using electronic answering machines on the phone system, allowing the caller to leave detailed messages for the recipient, and saving clerical and white-collar labor costs. This is significant because even the phone companies' own estimates suggest that three of every four phone calls do not reach the recipient. Answering machines cut down the number of calls and improve the quality of communication, but voice mail allows forwarding and sharing of recorded messages just like electronic mail, saving time over sending repetitive messages among various parts of the company.

To develop an understanding of the manufacturing, design, or any other process requires the knowledge and availability of statistical quality control (SQC) and statistical quality analysis (SQA) tools. Installing them on the same network with the e-mail system will enable all team members to access, use, and share all of the same data and the same quantitative analysis tools. Collecting the data is the first step, as shown in Figure 12.4. Note that the data from several batches has been collected across several tests. This data is typical of advanced manual data collection in the past, but this presentation does not help one to discover trends, or cause and effect.

The data analyzed by failure and arranged in a simple Pareto chart quickly points out where the largest problems are. An example of that analysis is shown in Figure 12.5. In the example, board design BF2HGI clearly has the most defects (38%) and the coding of the defects shows that most of the problems are caused by bridges (approximately 95 of the 160 defects) in board BF2HGI.

The observer should note that the primary use of the original data collection is not merely to store it electronically; it is to put it in a place and index it so that it can be analyzed and so that anyone needing it can find and use the same data without the need to construct or analyze it. Just as graphical representation of data allows one to see new trends that are not always visible in a tabular listing, an instantly accessible, indexed electronic storage of this data is more useful than a printed version in a file cabinet, because it allows diverse processing and encourages wide usage by all who have access to the computer network. This is one of the best ways to encourage communication. From this data one can then develop experiments to understand the interaction of these parameters and develop optimal manufacturing process parameters.

Figure 12.6 shows the results of an experiment that plots actual data against a curve fit and develops optimal yield-versus-thickness process settings, without extensive prototyping or direct on-line production experimentation. Later, once the team has determined what critical parameters control a part's producibility, statistical process control (SPC) tools can be used to control the process based on these parameters. Note the menu in the upper right part of the screen and the "plain English" explanation, as well as the scale in the center. These features enable wide usage of the analysis tool because they make it easy for the user to run the model, understand the meaning of the results, and choose the next logical step. Heretofore, a statistician would have been required to collect, analyze, draw conclusions, and interpret the conclusions. These examples illustrate how these tools will improve the understanding of the parts performance in the manufacturing processes. As that understanding evolves, areas for simplification of the design-to-manufacturing process will emerge.

One of the tools to help simplify the manufacturing process is group

SAMPLE YIELD ANALYSIS

CODE	TIME	CYCLES	BATCH #1	BATCH #2	BATCH #3	BATCH #4	BATCH #5	BATCH #6	BATCH #7
1 N44	11:28:15	218	0.493	0.438	0.502	0.384	0.463	0.507	0.478
2 B44	11:28:20	309	0.473	0.421	0.482	0.369	0.444	0.487	0.459
3 S14	11:28:25	662	0.404	0.359	0.412	0.315	0.380	0.416	0.392
4 B14	11:28:30	512	0.453	0.403	0.462	0.353	0.426	0.467	0.440
5 N02	11:28:35	114	0.581	0.517	0.593	0.453	0.546	0.599	0.564
6 B02	11:28:40	319	0.483	0.430	0.492	0.376	0.454	0.497	0.468
7 N22	11:28:45	404	0.463	0.412	0.472	0.361	0.435	0.477	0.449
8 B22	11:28:50	844	0.325	0.289	0.332	0.254	0.306	0.335	0.315
9 N24	11:28:55	56	0.729	0.649	0.744	0.569	0.685	0.751	0.707
10 N01	11:29:00	342	0.473	0.421	0.482	0.369	0.444	0.487	0.459
11 S17	11:29:05	94	0.690	0.614	0.703	0.538	0.648	0.710	0.669
12 B17	11:29:10	107	0.670	0.596	0.683	0.522	0.630	0.690	0.650
13 S22	11:29:15	255	0.591	0.526	0.603	0.461	0.556	0.609	0.573
#									

Figure 12.4. Data collection. (*Courtesy of BBN Software Products Co., Cambridge, Mass.*)

INFORMATION TECHNOLOGY REQUIREMENTS FOR DFM

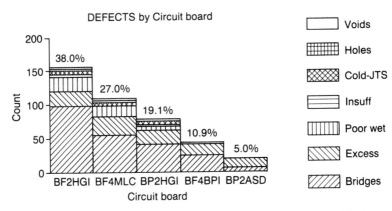

Figure 12.5. Pareto chart analysis. (*Courtesy of BBN Software Products Co., Cambridge, Mass.*)

technology: the technology used to classify and locate something by describing several of its common attributes. Rule-based producibility analysis tools, when linked with group technology, provide the team with the ability to evaluate design producibility, reduce part count, and lead to simplification of the shop floor. This offers fundamental advantages over databases as it allows one to correlate SQA data by part attribute, analyze the common attributes, group like manufacturing processes together to reduce floor space and travel, develop guidelines for manufacturability, and build up manufacturing process understanding by all who use the application.

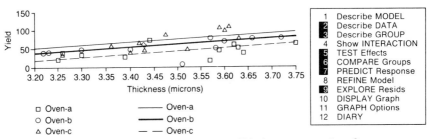

Figure 12.6. Data presentation to enhance utility and ease of use. (*Courtesy of BBN Software Products Co., Cambridge, Mass.*)

CIMTELLIGENCE GROUP TECHNOLOGY & CAPP SYSTEM						
Part Number CTC-0912.		Part Description POWER SUPPLY		Dwg Rev A	Pln Rev 2	Qty 12
Failure Rate 2.5	Vendor Rating 4	Rating Producibility Scale Rating 75%		Forecast Usage 200	Last Six Months Usage 170	
Next Assembly CTC-21921		Raw Material Type BILL OF MATERIAL		Material COMPONENT LIST		Cost $700.0
Diameter Length 8.0 Width 8.0 Height Thkness	Approval Eng. K. COLLINS Mfg. BOB GLATTER Prod. J. HYDRO Q/A. J. HARRIS			Date 07/08/90 07/19/90 07/25/90 07/30/90	Project A-22 CONVERTOR		Source H
^	^	^	^	^	Document Location On-LINE CADDS		
^	Design Specification/Remarks PER SPEC-1920			Mfg. Specifications/Remarks PER MIL STD-1567A		

Figure 12.7. Typical header display of group technology information. (*Courtesy of Cimtelligence Inc., Lexington, Mass.*)

Figure 12.7 shows a typical header display of group technology information for one part number. Note the associated dimensional and parametric data as one would expect from any current database. One should also note the failure rate, forecast usage, and producibility rating attribute data included in the display. This is a new association of useful data for two reasons: First, it allows data to be collected and referenced not only by part number but by manufacturing process attributes by all members of the DFM team—engineering, manufacturing, procurement, and quality assurance personnel. Before this, if the data had been collected, it was in various forms and resident in separate departments, causing data incompatibility and poor design choices, and delaying product development. Associating the physical as well as process attributes allows and promotes higher-quality decisions earlier in the design cycle. Second, because the data can be associated by attributes, manufacturability trends can be more easily observed and design decisions can be made more consistently. These capabilities enable the designer and the rest of the DFM team to see the effect of past design choices and build or revise manufacturability guidelines. This is the fundamental capturing of design rules and is a major step to improving the design cycle by promoting the learning process.

Figure 12.8 shows a display of rule-based producibility analysis, which was developed as a result of the use of SQA to understand which parameters

INFORMATION TECHNOLOGY REQUIREMENTS FOR DFM

1	Producibility Analysis
Analysis Name : Group Tech Classification Evaluation Category: PWB Power Supply	
Board Cost: $700	
QUALITY \# of Components : 60 Avg. Failure Rate/Component: 98.5	
COMPLEXITY \# of Layers : 7 Board Size : 8 × 8 Total Thickness : 80 Thickness Tolerance: 4 Producibility Total: 75% Mfg. Tol. Allowance (18–2 Mils) : 4 Inspection Sample Size (0–100%): 10	

Figure 12.8. Rule-based producibility analysis. (*Courtesy of Cimtelligence Inc., Lexington, Mass.*)

governed the manufacturability of the product. Then a set of relative scoring scales was developed for each parameter and the rules with their scoring scales were set up in the analyzer. Each design's parameters are loaded into the rule-based system and analyzed. The engineer can see the effect of his or her design choices on the producibility of the board. By developing a consistent quantitative scoring system, the engineer can see whether his choices are improving the design. Note for example, the average failure rate per component of 98.5 ppm. This data was developed from the collected data from the SQA system and loaded into the knowledge-based system. When the engineer selected the components for the design, he was able to collect the component failure rates and average them. This type of analysis can provide near-instantaneous feedback on the quality of design. By contrast, typical electrical engineering CAD (EECAD) systems today do not provide any scoring system. They only signal when the design is complete or is finished autorouting. Thus, most EECAD systems will provide a completed design, not an optimized design.

These group technology tools enable a systems engineering approach to problem solving by all members of the DFM team. While these tools are crucial to the initial DFM team, they can be expanded easily using the technologies below to form the basis of a company-wide DFM implementation.

Examples of specific enabling technologies required to support company-wide DFM include:

- Local and/or wide-area networks such as those complying with IEEE 802.3/4, TCP/IP, OSI, or MAP standards.
- Graphical user interfaces such as X Windows with Motif, an industry standard user interface that uses icons, windows, and presentation standards to promote consistent man–machine interfaces for computer applications. X Windows is an especially powerful tool since it also permits remote graphical usage of an application, instead of requiring costly local computer hardware at each user's location.
- POSIX compliant operating systems such as UNIX.
- SQL data bases such as Ingres, Informix, or Oracle.
- Off-the-shelf applications that run on the above:

    Solid modeling and simulation applications, such as Hewlett-Packard's ME 30, McDonnell Douglas' Unigraphics, Patran, and ANSYS.
    Statistical quality analysis, statistical process control, such as BBN's RS Series products.
    Group technology/computer-aided process planning (e.g., Cimtelligence's Intellicapp).
    Producibility analysis tools (e.g., Boothroyd/Dewhurst DFA, Hitachi, AEM).
    Configuration/release control and change management (e.g., Sherpa Corporation's Data Management System).

This list includes many standards-based technologies. They represent the most widely available technologies, providing the highest insurance against getting the DFM team side-tracked into developing their own technologies. They provide the highest investment protection against any vendor who attempts to control market share through proprietary technologies. The goal is to implement DFM, not create another inhibitor to information flow. Finally, these technologies also offer the safest approach to ensuring that other tools in the evolution of DFM will be in the standards-based technologies. It is simply too expensive to develop products in other environments. IBM's development of the personal computer emphasized standard modules, and in doing so, cut significant cost and time from the development cycle. Finally, the standards-based technologies can be installed and supported with the least risk of unique, custom (costly) integration efforts. As an example, Figure 12.9 shows a typical U.S. electronics company starting to implement DFM. Its information systems design incorporates both of the example applications. Data is collected from the shop floor and passed on the network to the analysis server for statistical processing. Once

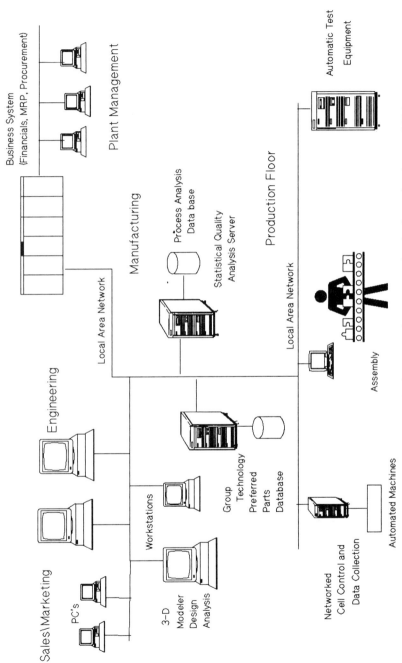

Figure 12.9. A typical U.S. electronics company beginning to implement DFM.

processed, it is sent over to the preferred parts database, which is indexed by the group technology server. Engineering uses the preferred parts database when designing new products and also uses it as an analysis server to evaluate new designs for producibility. Once the design is optimized, it is released to the floor via the MRP system. Procurement orders all parts and uses the input from the preferred parts database for administering the supplier quality assurance program.

12.3 PLANNING THE IMPLEMENTATION OF TECHNOLOGY TO SUPPORT DFM

Where does one begin? Since DFM's major contribution is to the ability to compete, the goals of the organization are the starting point for planning DFM information technology implementation. Management must establish and clearly communicate its vision and goals. Management's strategic business goals are the beginning point for establishing the team's metrics and priorities and expected payback. These goals need to be more fully articulated than "Make more profit" or "Improve product quality." They need to be quantifiable, and apply across the DFM team, if not the whole company. John Young, President of Hewlett-Packard Co., believing that improving product quality was the key to successfully competing in the 1980s, established the 10X Quality Improvement goal at Hewlett-Packard. He stated that he wanted a tenfold increase in product quality in 10 years. He left the method of implementation to his general managers' discretion, but he defined a clear goal that became a common benchmark for the whole corporation. In each subsequent meeting, progress toward this goal was reviewed and corrective steps were identified where needed. During these 10 years, HP divisions developed the metrics, organization, methods, and technologies to meet this goal. Ten years later, HP has met that goal and continues today to build on its experience in producing high-quality products.

For companies considering DFM, the first step is to make time-to-market reduction one of the company's strategic goals. If management is committed to a smoothly running organization, then time-to-market reduction will be high and visible on its tactical and strategic list of goals. Emphasizing product quality or fast delivery is not definitive enough. Cutting the time to market by 25% over 2 years is a definitive goal. With that goal in place, the DFM team can measure its contribution, progress, and investment justification.

To develop an information technology implementation that provides a vehicle for supporting DFM's goals, one needs to develop a supporting goal for information technology's role in DFM. With respect to this discussion, the goals of DFM that information technology supports are to:

PLANNING THE IMPLEMENTATION OF TECHNOLOGY TO SUPPORT DFM   297

- Provide the means to ensure continuous, consistent feedback to all members of the product team (marketing through field service) in time to avoid delay, cost, or quality impacts.
- Enable empowered employees to make real-time decisions with accurate estimates of the cost, product definition, time to market, and quality trade-offs.

Note that there is the potential for discordant goals between DFM and computer-integrated manufacturing (CIM) especially if the firm's drive toward CIM is focused on automation. This can be seen as forcing more restrictive rules on the design team and inducing a lack of flexibility in the manufacturing process. A human worker is far more adaptable to a new design than even the most advanced flexible manufacturing system. But the fact is that shortening design lead time, not manufacturing cycle time, is the critical tactic in developing flexibility. Flexibility in manufacturing is not the ability to build whatever engineering designs, regardless of the cost or difficulty; it is the ability to change product mix and volume, and manage new product introduction cost-effectively to meet changing market demands. While the first two are governed by reduction of setup or changeover time, the latter is governed by the ability to bring designs into manufacturing rapidly, without redesign. By concentrating efforts on bringing the most producible design into manufacturing rapidly, a company builds flexibility into its entire business. The Japanese have known this for almost a decade; in a 1984 survey of Japanese chief executive officers (CEOs), twice as many said they were putting efforts into improving their design-to-manufacturing cycle over any other area (Kaplan, 1990).

Independent research by McKinsey, Gomory, Hayes, Peters, Bower, and others points to several critical steps that companies must take to reduce time to market in order to compete effectively in the 1990s. Each author has his own point of view, but the consensus of their viewpoints yields these critical steps: Management has a crucial role in defining DFM's goals; building cross-functional teams is essential to success; and enabling workers through horizontal information systems and teams is a proven method of improving product quality and reducing costs.

12.3.1 Management's Role in DFM

As already noted, management's major contribution is in making time-to-market reduction a critical goal. In each meeting, management should check the progress against the goal through metrics developed by the DFM team. Management should require marketing to identify new trends and their appropriate metrics to respond to changing customer demand.

These goals and employees' understanding of them are critical to DFM's success. They govern the charter of the DFM team, form the basis of what and how the process will be measured, and become the measurement criteria for the cost justification of any investment to support the DFM project. Thus, management needs to communicate those strategic goals clearly to all members of the company. This is leadership: establishing a vision and inspiring team members to accomplish the goal through clear communication to everyone. For more detail on goals and metrics, see the section on metrics in Chapter 5.

Last, management must allocate funding to support the investments in teaming, training, tools, and information technology. It will seem a hollow message if the first two activities are taken on without the follow-through of this activity. Good followers take on goals, but require the leadership and support to stick with it.

12.3.2 Identify the DFM Implementation Team

In implementing DFM, a team must be established with the charter, the authority, the skills, and the tools. These people will be the role models for change. To successfully change the organization, the team needs to be supported through clear communication and tangible support. The message to the rest of the organization should be "We reward risk takers. These people are building our company to compete successfully and we back them 100%."

In terms of specific actions this means:

- Organizing the DFM teams with the authority (product and process budget) and responsibility (product/market performance) to succeed. Marketing, procurement, R&D, manufacturing, engineering, components engineering, documentation, and field support should participate in DFM teams.
- Using a structured methodology and deciding on design goals that include cost, schedule, and quality as well as feature set to design products and achieve consensus at predetermined milestones before proceeding to the next phase.
- Enable the team with budget and tools to ensure ease of information flow.

While it may seem early in the process, this is the appropriate time to select strategic vendors for implementing DFM. Some of the critical vendors (e.g., networking, quality, and group technology tool vendors) should be brought in here; their experience in other projects will provide a sounding board and insight into achievable metrics and returns. Also, the more a

PLANNING THE IMPLEMENTATION OF TECHNOLOGY TO SUPPORT DFM 299

vendor understands the project the better can he support the team's needs. There is risk in the vendor who states his firm "can do it all" or claims that his entire family of products are just what the company needs. No firm can truly do it all, and the only case where the company needs an entire vendor's product line is probably a brand new plant, a "green-field project," where the vendor has a unique offering. This case will be obvious, and is extremely rare.

12.3.3 Define Metrics

Metrics provide the team with measurements of progress and the basis for decision making, either in product design or in information technology investments.

The first step is to set measurements and milestones for each project; hold regular participatory management reviews and measure cycle times across the enterprise and across departments. Tracking the quality of each stage's output is important to improving the cycle time. Suitable measures include the number of rework cycles and percentage of rework at each stage of product development. These provide ongoing measures of progress toward development cycle reduction.

The team should begin by instituting a metrics team to develop the first set of cycle metrics. The metrics team should be a subset of the DFM team and their charter should be to establish some initial benchmark(s) on design-to-shipment cycles and other metrics. Superior metrics are simple ones tailored to the design and production environment; expect several iterations before settling on the metrics that best fit the company's processes and environment. Since time is the critical resource, a good metric will usually measure work per unit of time normalized for the level of complexity. As these measurements are used, it will become clear where to broaden scope, reduce redundancy, and change the organizational structure to be more responsive to the market before product release.

12.3.4 Define Critical Functions and Information to Support Those Functions

In choosing and implementing metrics, it will become clear how product design, material, and information currently flow through the organization. Develop this view into a diagram showing all three flows and include the critical functions and information transfers as a product moves from concept to shipment. Use a systems engineering (top-down) approach. The first sketch should have no more than 6–10 functions (e.g., production release) and 10–20 information transfers (e.g., an arrow showing as-built configuration being passed up from the shop floor to the knowledge base). The diagram should be a flow chart of activities across the enterprise in the idea-to-manufacturing cycle. Figure 12.10 shows an example of a typical electronics company's information flow.

Sections can be expanded later to include specific information as needed for system implementation. Be sure to identify each source and destination of every design, material, and information flow shown. Be sure that the diagram represents a functional flow, not a "whose department or computer has it now?" approach. Figure 12.11 shows an example of a detailed flow through an electrical engineering department of an electronics system manufacturer: Note the individual functions, flows, and measures of work per unit of time.

Next, the team should annotate the flow with the length of time and manpower each step takes for a single full cycle. The team should use estimates or available data at first; later, measure these estimates. There are many ways to measure these activities (Holden, 1986), but the most important thing to remember is to make the measurement consistent across the process. If any step is repeated, draw that in and estimate the time and manpower it consumes during that repeated step. This graphic representation of the flows through the organization, when developed and agreed upon by the team, becomes the roadmap to implementing DFM information infrastructure. There are several tools that can help to develop this graphic map of flows in the organization, such as flow charting, structured analysis, and statistical analysis tools.

Using statistical quality analysis tools, such as the Pareto chart (e.g., Figure 12.5), identify the function that takes the longest. Examine that function to determine which aspect is responsible for the greatest amount of time. In addition, other statistical quality analysis tools such as the Ishikawa Cause and Effect Diagram (Ishikawa, 1985) (also known as fish-bone diagrams), may help identify some of the more obvious contributors to the time taken by the longest function.

Using these tools, it is possible to develop problem-solving techniques to determine where to simplify processes, where most data is needed, where to automate, and what to integrate. These analyses will enable the development of ideal flow, and characterize its length and required manpower. This will not be a single, one-time exercise but a continuous process, as business and available technology change.

Once the longest event and its cause are identified, the team can develop a plan to reduce it. In selecting information technology solutions, be careful of using existing systems to move data; they can convolute or add extra translation steps without adding real information value beyond formatting. For example, when looking at downloading numerical control (NC) instructions

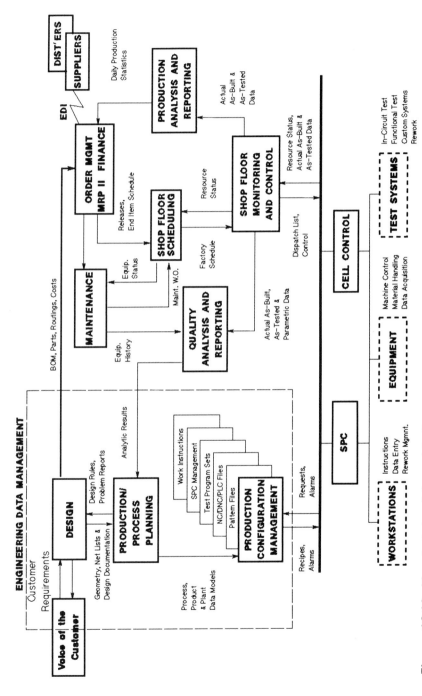

Figure 12.10. Manufacturing information flow in a typical electronics company. (*Courtesy of Hewlett-Packard Co., Palo Alto, Calif.*)

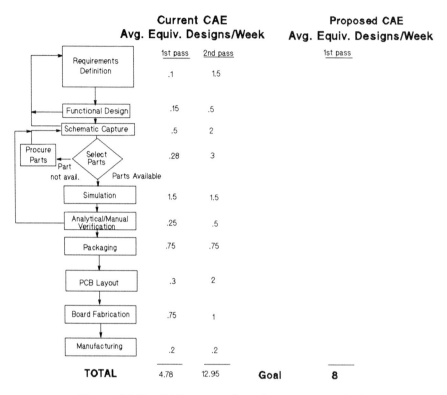

Figure 12.11. CAE-to-manufacturing process analysis.

to a machine, it is typical to select an internally developed PC-based solution to cut the cost of the hardware. However, this solution seldom accounts for the added development cost, or the added scrap cost of downloading the wrong instruction at the wrong time. By contrast, a commercially available application that is integrated with the shop-floor control system to synchronize the downloading with the shop orders costs more up front but provides greater benefits because it reduces both time and errors.

To illustrate this, Box 12.1 shows a typical example of the relative costs of providing an NC solution for one machine for one year. In the example, the internally developed system does not integrate with the shop-floor control system; engineers actually spend more time deconflicting the internal solution with the rest of the shop floor. The cost of scrap material is estimated to be approximately 0.2% of the revenues of a $200 million company, and it is

BOX 12.1: TYPICAL COSTS FOR NC SOLUTION

Savings category	Status quo	Internal development	Commercial application
Labor			
Direct	2 hr/mth	$ 288	$ 288
Manufacturing overhead	4 hr/mth	1,440	2,880
Engineering overhead	5 hr/mth	−1,000	3,000
Material			
Scrap	$30,000	10,000	28,000
Total	0	10,728	34,168
Solution costs			
Hardware		7,000	30,000
Networks		2,500	4,500
Development labor		125,000	15,000
Software		1,000	35,000
Total		135,500	84,500
Net savings (first year)		124,772	50,332

one-third worker-caused, two-thirds machine-caused. Additionally, the cost of development is 1.25 salaried engineers and their overhead for one year. Support costs are not shown; they would show similar results.

As one can see, in the internally developed solution, direct labor shows a very small saving, engineering overhead actually increases because of additional work and software maintenance. The net effect is that the internal solution takes 4–5 times longer to pay off, while the commercially available application pays off within 3 years. This brief example also does not take into account that the internal solution takes 1 year to develop, while the commercially available solution is usually implemented within 3 months and saving the company money within 6 months. In conclusion, one should be very wary of the solution using internally developed "homegrown" solutions. They should be examined to ensure that they truly provide integration, and do not add non-value-added steps to the engineering-to-manufacturing process.

12.3.5 Estimate Costs and Paybacks

This section will discuss costs and cost justification for information technology to support DFM. Specifically, when looking at investment justification, one should ask questions such as:

By providing this information, where will the DFM team save manpower in the business?
Where and how will scrap be reduced?
How much will the quality of the product be improved?
How will the company be able to respond to a given order or to a change in an order more quickly?
How will this investment enable the company to respond to a change in the market demands?

Using this information, how can the organization structure or business practices be streamlined to reduce the cost of product development, production, and support? What other investments must be made to ensure the success of this investment? In every answer, companies will need to identify how the savings that occur provide the source for cost justification. Further, since the team developed the functional flow diagram, the diagrams can be used to estimate reduction in time and manpower in the changed process. In essence, as Tom Peters describes it, one is re-engineering the business, not for the sake of technology, but because it allows one to do things more competitively (with less cost and in less time) than one could before.

It is also important to identify the costs associated with the investment. In traditional cost-accounting, only the hardware, software, and direct labor hours would be included. In this investment analysis, just as all sources of payback will be included, total life cycle costs will also need to be included, such as maintenance and indirect or support staff.

Numerous methods have been proposed to justify capital investments; most have been based on return on investment (ROI) (Clark, 1982), return on assets (ROA) (Vollmann et al. 1988), or discounted cash flow (DCF) analysis (Kaplan, 1986). These methods are derived from accounting structures that serve to develop costs and returns for reports to the stockholders. Research on these methods shows that they fail to recognize what are often called the "soft benefits" or "opportunity costs." This means that these commonly used investment justification methods fail to recognize the greatest benefits of a DFM program! Another method, Hurdle rate analysis is based on comparing project returns against the going rate of return on capital. It induces more and more conservative assessment of returns, inhibiting investments. Unfortunately, none of these methods measures the strategic value of any investment. This is especially tragic, as "40% of a manufacturer's

PLANNING THE IMPLEMENTATION OF TECHNOLOGY TO SUPPORT DFM 305

competitive advantage comes from major injections of new technology, both equipment and processing techniques," according to W. C. Skinner (Pennar, 1988, p. 104). Other studies have shown that design determines 75% of the product cost before the design reaches manufacturing. More recent quantitative measurements measure total factor productivity (Hayes et al., 1988, pp. 378–389), and its impact on profitability. This is a tremendous improvement since it shows the impact of productivity on profitability—the impact of inefficient design processes on the company's profitability. But even this measure has difficulty in discerning the contribution of new technology investment versus the more efficient use of existing technology.

Currently, according to Hayes, Wheelwright, and Clark, the most productive cost justification takes into account three areas (Hayes et al., 1988, pp. 93–94):

> ... how the new capability adds value, the cost of implementation, and the difficulty competitors face in implementing it. It consists of understanding how [DFM] might interact [with other capabilities] to create additional value... Second... is an analysis of what the investment means for the financial health... of the enterprise. The third element is... an analysis... of how the investment... fits with the strategic directions of the business and its major functional areas.

This analysis can only be accomplished with an understanding of the goals and business strategy that were discussed earlier.

While each company has its own justification process and format, Box 12.2 is an example of an economic justification form that addresses these three areas and can be used for DFM investment justification.

12.3.6 Gaining Management Approval for the Project

By this point, it is assumed that management has clearly defined its goals. Each company has its own approval process. However, almost all of them require a complete presentation of the scope, costs, impact, savings, and schedule. The development of such a proposal that has evolved successfully usually contains the following:

1. The DFM team agrees on an "as is" and "to be" graphic description of the organization's flows (as discussed earlier).
2. The DFM team agrees to the metrics and the methods of measurement.
3. The DFM team draws up the economic justification, each representative contributing their costs and savings from their department.
4. The DFM team selects a spokesperson who represents all members.
5. A proposal is developed and presented. It can consist of the following:
    - Purpose and agenda of the meeting.
    - Goals of the DFM project: Identify how the goals support the business strategy.

BOX 12.2: STRATEGIC INVESTMENT JUSTIFICATION

Section One

Estimate savings: Describe and quantify the savings to be accomplished by this investment. Include estimates of the impact of this investment on employee morale and productivity. Use burden rates from the controller's office for labor costs. Attach additional forms or diagrams as necessary.

		Value of
Area of savings	How savings occur	savings

Total savings

Opportunity costs: Identify the potential additional expense the company would incur if this proposal was not accepted (include an estimate of the cost of the status quo).

Total opportunity costs:

Total savings if investment is approved:

Section Two

Investment analysis: Using the comptroller's office's cost-accounting methods, calculate the return on assets of the proposed investment. Include all external and internal costs (such as additional staffing). Identify any costs or savings not included in the analysis. Identify all assumptions and estimates as such.

Section Three

Strategic goal accomplishment: Explain how this investment supports the company's strategic goals. Be explicit and as specific as you need to be.

SUBMITTED BY: DATE:

REVIEWED BY: DATE:

APPROVED BY: DATE:

- Estimates of benefits and sources of return.
- Scope and description of the project.
- Estimated cost and DCF or ROA (as required by the company): Identify any estimates, unknowns, and risks.
- Implementation schedule.

6. It is essential that all members of the team support the presentation; prolonged dissention without compromise only serves to defeat the best efforts of the team.

12.3.7 Detailed Implementation Planning of the Pilot Project

Once goals have been set, the justification has been characterized, and management has approved, the detailed implementation work must be planned. This is where most projects meet success or failure, because this is where the lack of clear definition of either goals or scope first meet reality.

Use project management and structured analysis tools to define the exact scope of the pilot down to the data elements, participating organizations, users, support requirements, and schedule. Identify any unknowns, and resolve them. If forced to use assumptions, post them so that if any of them are in error they can be readily corrected. Identify the first group of users and involve them in identifying exact data elements and how they should be presented to them. Train them on the application concepts. Creating a positive environment for change assists in both the ease of implementation and the results.

12.4 IMPLEMENTING DFM TECHNOLOGY

12.4.1 Selection

There are many guidelines available for selection of information technology. Regardless of the application, information system, or piece of DFM methodology one is implementing, there are some common ideas that should be considered. These are important because, in the past, many companies have selected vendors' products primarily on the basis of the product's technology. The successful contribution of the product, as they later learned, was dependent more on the ease of use of the human interface, implementation, or support. As an example, Apple Computers' appeal has been attributed to the ease of use of the human interface. Figure 12.12 shows some common criteria and means of measuring a vendor's qualifications.

	A	B	C	D	E	F	G	H	I	J	K
1								Scale	Raw	Total	
2	Criteria:			Measurement:				Factor	Score	Score	Comments
3											
4	Company market presence			% market share				1		0	
5	Company commitment			Product plans clarity				2		0	
6										0	
7	Availability of sales			No. of sales, support personnel				1		0	
8	personnel			available							
9	Knowledge level			Avg. years Application experience				2		0	
10	Vendor's Added Value			Understanding\Contribution to the						0	
11				Project Design or implementation				3		0	
12	Feature/functionality			% coverage of required functions				1		0	
13	Ease of use			Est. time to productive use				3		0	
14	Price/Performance			% functionality coverage x cpu powr				2		0	
15				/hardware & software cost						0	
16	Installation			Costs and Time to install				1		0	
17	Integration costs			Costs to link to applications				3		0	
18	Maintenance support			Lifecycle cost of vendor maint. and				2		0	
19	for h\w and s\w(5 yr.)			any add'l company staffing						0	
20										0	
21	Concept education			Availability & cost\student of				1		0	
22				concept education						0	
23	Product operation			Availability and cost\student				3		0	
24	training			operator training						0	
25										0	
26	Vendor's Business stds			Communication effectivity measured						0	
27	and ethics			by response quality and speed				3		0	
28	Other:									0	
29										0	
30										0	
31	Total:									0	
32											
33	Note: Scale Factor:			1 = normal importance		3 = critical importance					

Figure 12.12. A typical vendor evaluation.

308

IMPLEMENTING DFM TECHNOLOGY

Successful implementations have shown that heavier weighting on support and integration costs over feature/functionality is critical to good vendor evaluations. The best evaluations also include an assessment of the culture of the vendor; it is not good business if the firm and the vendor cannot communicate.

12.4.2 Procurement

Discuss the procurement process, approval chain and cycle, standards of pre- and post-purchase service performances, and project schedule with the vendor up-front. In the past, many companies have regarded this kind of information as confidential. This confidentiality resulted in a poor understanding of a company's needs, and invariably it showed up in disappointed teams, poor information system performance, and delays in the DFM project schedule. Sharing this information with the strategic vendors will provide the vehicle for improved understanding and more timely and correct service, and provide the vehicle for best practices sharing on both companies' part. Both the company and the vendor will benefit from the exchange by gaining a clearer view of the other's offering, methods, and requirements.

12.4.3 Training

Training is becoming one of the strategic differentiators in helping a company compete in the 1990s. With the rate of change noted above, and recent federal research that suggests the skill levels of Americans are declining, training is becoming more important than ever before. This includes both concept education (e.g., "What are the robust methods of design concepts and how can they be used in manufacturing electronic products?") and product feature operation training (i.e., "What button do I push to get control charts?"). The latter type of training is offered by almost all information system or application companies today. However, this requirement should decline as the growth of applications with near-intuitive user interfaces (e.g., X-Windows with Motif) decrease the need for operator training. Conversely, the need for concept education will grow as the existing work force needs retraining, and the need for in-depth understanding of these concepts grows. In a continuous improvement cycle, workers will need to have an education path of continuous improvement to reinforce the learning process.

12.4.4 Integration

There has been much already written about the needs and benefits of integrating one's business; justification for such will not be attempted here.

Integrating information systems depends a great deal on the planning and design of the integration. As noted earlier, if the information is passed through several existing computer systems, it is easy to end up with several information reformatting and translation steps without adding real value to the information itself. Therefore, planning of integration is critical because this function can either inhibit or promote ease of communication by the choice of technologies, vendors, and design of the network(s). With respect to the topic of integration, it is important to clarify several steps, starting with the simplest.

Configuring the system for the number of users, applications, peripherals, and so on should require no custom code and, as such, is not systems integration. *Systems integration* is the process of causing two dissimilar functions or applications and/or systems to communicate information, upon user request. Since this is an area where most definitions are hazy and most deliverables are paper designs until it is too late, it is important to assess the software engineering skill levels of any proposed systems integrator. It should not be confused with *custom development*, where the vendor or the observer develop the application with a "tool set" on top of an "environment" or database. This will usually get the observer a unique, stand-alone (soon to be unsupportable) solution and a lifetime support contract with a single vendor without competition. When embarking on an integration effort, a company should use as many off-the-shelf tools and standards-based technologies as possible, to avoid being snarled in custom code and a lifetime contract with no competition. While each company's data values are going to be unique, one should be able to procure applications with a standard import and export data facility that can be used to pass data files until the company is sure of its integration needs. Any application without such features is probably not worth the investment.

12.4.5 Support

Historically, support after installation has been a justification for data processing staffs. As companies look for areas to cut cost, they are out-sourcing this type of support because a support contract is usually less costly than maintaining the expertise and resources to handle all maintenance within the company. When evaluating vendors for support, using some of the criteria and measurements shown earlier will provide an objective basis for selection and more realistic expectations on the part of both company and vendor.

Another area of support that is growing is the role of internal systems developer. In this role, the data processing staff, as members of the DFM team, act as the information system project coordinator, handling the scheduling and logistic details of the information system(s) implementation.

When this is done as part of the team, and all members participate in the system selection, better systems and more successful projects result. All parties recognize the value of the system and the constraints under which it must operate, and the essential service it must perform. Further, the more knowledgeable are the users of the system, the more likely will they be to use it to the fullest extent, thereby getting full value for the company in its investment.

12.5 LESSONS LEARNED

One of the most valuable lessons to be learned is that technology cannot drive DFM. DFM is primarily a practice that uses technologies, either applied during understanding and characterization of a process, or as a means of improving or automating a stable process. Thus, the applied technologies provide information only as a catalyst for learning in an organization dedicated to continuous improvement. Technology applied in lieu of training or learning usually falls short of its expected payback.

Building on the successful investment of this technology, where can the DFM team expect to go? Expanding the use of these principles beyond the interaction of engineering and manufacturing to include marketing, procurement, and field service will eventually involve the entire company. This is the beginnings of making CIM a reality. CIM has been discussed widely, and it is the technological next step in developing world-class manufacturing.

First, what are the relationships between information technology, CIM, and DFM? Fundamentally, DFM is a building block to achieving CIM, as it provides the understanding of manufacturing and the integration with engineering. As DFM begins to build understanding of manufacturing, it provides a methodology for continuous improvement and begins to integrate some of the practices and information infrastructure, leading toward a more cohesive, competitive organization. Unfortunately, this is not a new lesson. Any discussion of DFM and CIM requires definition of the concepts that will be used. Each time these terms have been used, they have been defined from the perspective of the writer. As a result, the concepts have varied widely in their meaning and scope. For example, the American Production and Inventory Control Society defines CIM as (Wallace and Dougherty, 1987):

> The application of a computer to bridge and connect various computerized systems and connect them into a coherent, integrated whole.

A broader definition is provided by C. H. Fine in *Strategic Manufacturing* (Moody, 1990):

> [CIM] is the use of computer technology to link together all functions related to the manufacture of a product.

This definition begins to recognize and account for the business strategies, organization structure, business practices and/or procedures, and technologies that are all necessary to running a business successfully. For the purposes of this discussion, the latter, more encompassing definition will be used.

Additionally, factor automation and flexible manufacturing systems are two concepts that one should understand. They are often confused with CIM. In the case of factory automation and flexible manufacturing systems, the Japanese definitions offer the most concise view of the concepts. The definition of FA approximates the technological scope of our CIM definition (Kaplan, 1990):

> FA can be defined as the automation of the factory through the usage of a flexible manufacturing system (FMS), computer-aided design/computer-aided manufacturing (CAD/CAM), and office automation. Among these three basic elements, FMS is the core of FA. Typically designed for mid-range volume and mid-range variety production, FMS integrates industrial robots, numerical control (NC) machines and automated materials-handling systems using the concept of cellular manufacturing.

CIM goals have a great deal in common with DFM, in part because of the scope of our CIM definition. But even if the more restrictive definition is used, DFM is a building block to achieving CIM, as it provides the understanding of manufacturing and the integration with engineering, two quid pro quo's to automating the manufacturing process. DFM provides the basis for process integration, the fundamental goal of CIM, and CIM provides the enterprise-wide scope to expand DFM practices to the rest of the company.

12.6 CONCLUSION

In summary, market forces are making time to market a critical measure of success. DFM provides a methodology for understanding and reducing the components of the company that drive development time. Information technology's role is changing to provide the critical cross-functional flow of information across the company. This information flow enables constantly improving decision making—a sustained learning process that is a trait of successful companies in the 1990s. Since U.S. manufacturing has developed efficient mass-production methods, the efficiency of this information flow is the gating factor in determining time to market; its efficacy determines the response to customers.

In evaluating technologies to support DFM, flexibility, fast response, consistent easy usability, wide access, cost-effectiveness, and availability are the most important attributes to emphasize. Information technology implementation that accentuates these attributes will promote the use of DFM methodology, not inhibit it. In this new role, integration and usability, not pure feature/functionality performance, are the major attributes that ensure success. This integration challenge is enormous since most companies have already made some investment in materials management, financial, and CAD systems.

The steps to planning DFM information system implementation include: management development of goals and vision; development of a team with the authority (budget) and responsibility (e.g., measurements against the goals above); definition of metrics of the design cycle; definition of the information flows, including destinations and sources; and finally, development of the costs and paybacks to show how the investment will support the company's goals. One should be open to looking for benefits throughout the company, as current accounting practices do not recognize most of the opportunity cost benefits that a DFM implementation will provide.

The steps to implementing DFM information technology include: detailed data and analysis flow and implementation design; implementation and integration; training, start up assistance and phase-in of support. During implementation, vendor selection, procurement, installation, configuration, integration, training, and start-up support are planned and executed. Vendor selection mirrors the DFM information technology requirements: It is the reliability and responsiveness of the vendor, not the technology that should take priority in selection. Measuring progress and results is important because the results give the management and the DFM team the first step to understanding the next step to take.

Finally, DFM is a methodology that provides a crucial stepping-stone toward integrating engineering with manufacturing, a necessary step to achieving CIM and a critical step to being a quality-, cost-, and delivery-competitive company in the 1990s.

REFERENCES AND SUGGESTED READING

Berliner, C. and Brinson, J.A. (eds.) *Cost Management for Today's Advanced Manufacturing: The CAM-I Conceptual Design.* Boston: Harvard Business School Press, 1988.

Bower, J.C. and Hout, T.M. "Fast cycle capability for competitive power." *Harvard Business Review*, Nov. 1988.

Clark, J.T. "Selling top management—understanding the financial impact of manufacturing systems." Paper presented at the annual APICS Conference, 1982.

Clark, Kim B. "What strategy can do for technology." *Harvard Business Review*, Vol. 67, No. 6, 1989.

Drucker, Peter. "The emerging theory of manufacturing." *Harvard Business Review*, Vol. 68, No. 3, pp. 97–98, 1990.

Freund, M. "Why you need to understand management's view of technology." *LAN Technology*, p. 20, June 1990.

Gomory, Ralph E. "From the ladder of science to the product development cycle." *Harvard Business Review*, Vol. 67, No. 6, p. 100, 1989.

Hayes, R.H., Wheelwright, S.C. and Clark, K.B. *Dynamic Manufacturing*. New York: Free Press, 1988.

Hayes, R.H. and Jaikumar, R. "Manufacturing's crisis: new technologies, obsolete organizations." *Harvard Business Review*, p. 77, Sept. 1988.

Hewlett-Packard Co. *Multivendor Network Management*. Palo Alto, Calif.: Hewlett-Packard, Co., 1989.

Holden, Happy. "Six principles of automation in computer integrated manufacturing." Paper presented at 1986 Hewlett-Packard CIM Seminar, Cupertino, Calif., May 1986.

Ishikawa, K. *What is Total Quality Control? The Japanese Way*. (Trans. D. J. Lu.) Englewood Cliffs, N.J.: Prentice-Hall, 1985.

Kaplan, R.S. "Must CIM be justified on faith alone?" *Harvard Business Review*, March 1986.

Kaplan, R.S. (ed.) *Measures for Manufacturing Excellence*, p. 40. Boston, Mass.: Harvard Business School Press, 1990.

Lipsey, R. and Meyerson, M. *The Landmark MIT Study: Management in the 1990s*. Boston, Mass.: Arthur Young Co., 1989.

Miller, Jeffrey, G. and Roth, Aleda V. *Manufacturing Strategies*. North American Manufacturing Futures Survey. Boston University, 1988.

Moody, Patricia (ed.) *Strategic Manufacturing*, pp. 264–265. Homewood, Ill.: Dow-Jones Irwin, 1990.

Nevens, T.M., Summe, G.L. and Uttal, B. "Commercializing technology: What the best companies do." *Harvard Business Review*, Vol. 68, No. 3.

Pennar, K. "The productivity paradox." *Business Week*, p. 102, June 6, 1988.

Peters, Tom. *Thriving on Chaos*. New York: Harper & Row, 1987.

Schroeder, M. and Gross, N. "How supercomputers can be super savers." *Business Week*, No. 3181, 140B, 1990.

Stalk, George, Jr. "Time—The next source of competitive advantage." *Harvard Business Review*, Vol. 66, No. 4, 1988.

Vollman, T.E., Berry, W.L. and Whybark, D.C. *Manufacturing Planning and Control Systems*, p. 30. Homewood, Ill.: Dow Jones-Irwin, 1988.

Wallace, T.F. and Dougherty, J.R. *APICS Dictionary*, 6th edn., p. 6. American Production and Inventory Control Society, 1987.

Whitney, Daniel, E. "Manufacturing by design." *Harvard Business Review*, Vol. 66, No. 4, 1988.

CHAPTER 13

Knowledge-Based Engineering

The first computer-aided engineering systems were actually automated drafting systems used for generating design drawings. Later, computer-aided manufacturing systems automated the generation of numerical-control output for machine tools. Increased quality, shorter development cycle times, and lower costs are crucial to ensure manufacturability and to bring products to market sooner. These CAD/CAM systems did not address the area with the greatest potential for reducing development time and costs—which is design (Potter, 1989).

13.1 LIMITATIONS OF TRADITIONAL CAD SYSTEMS

Traditional CAD programs (Figure 13.1) are essentially interactive drafting tools and are geometry-driven; that is, the definition of some model within the program begins with the geometry. The CAD system is used to manipulate the geometry until the exact set of points, lines, and arcs completely defines the geometric form of the design. The limited geometric information developed on a CAD system is insufficient to drive the rest of the product-engineering process, including activities such as process planning and Material Requirements Planning (MRP). "In effect, this is not design—making choices about the shapes, sizes, and relationships of parts—but merely documenting a design that came from elsewhere" (Robinson, 1989).

Design choices involve constraints, rules, or equations that relate one parameter of the model to another. For example, the diameter of a cavity insert might be constrained to:

```
{13.1}  (+  (the :part-design :diameter)
        (twice
          (the   :insert   :wall-thickness))).
```

Similarly, a mold plate thickness may be constrained by the injection pressure of the injection molding machine; and if the injection pressure is changed,

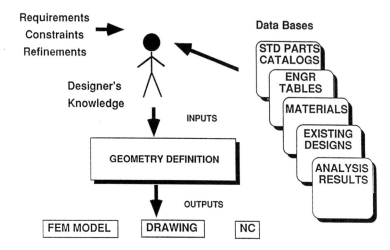

Figure 13.1. Traditional design process.

the mold plate must "inherit" this information otherwise the plate may undergo unsatisfactory deflection. Traditional CAD systems do not capture any of this information in the model. Geometry on a CAD system only defines portions of geometric entities. It incorporates no understanding of the engineering relationships that determine these entities (Robinson, 1989).

13.1.1 Adding Intelligence to Objects

If objects contained knowledge of their behavior and relationships to other objects in their application environment, they would be able to "design themselves." The creation of parametric descriptions of objectives involves capturing constraints and behavioral rules of the objects and their response to input values. The capability of modeling such behaviors will make possible the construction of models able to generate a viable design given some performance criteria (Belzer and Rosenfeld, 1987).

Knowledge-based systems solve this problem by incorporating the design rules or constraints into the model. Knowledge-based systems use a hierarchical structure to let the user define relationships among parts and assemblies. Modification of any part of the model will automatically reflect changes in any related part. In the previous example, change the injection pressure and the mold plate changes too. The user can construct a comprehensive but flexible product definition incorporating the knowledge that determines the geometric representation of the model as well as the ability to automate repetitive engineering tasks and integrate manufacturing expertise. It includes all the relationships among parts, not just their locations.

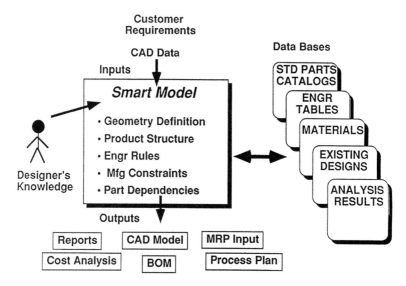

Figure 13.2. Knowledge-based engineering.

Associative Geometry

Knowledge-based, object-oriented systems allow full integration of geometric and nongeometric information in the same model (Figure 13.2). This associative geometry capability is an important feature of object-oriented systems. It allows geometric relationships between any geometric entities and allows the construction of geometric constraints. The nongeometric, or attribute, information, along with the geometric information, fully describes the object's properties and spatial relationships forming the complete "intelligent" model of the object, thus capturing the design intent.

DEFINITION OF OBJECT MODEL

An object has two components that comprise its total definition—geometric and nongeometric, or attribute, information.

GEOMETRIC INFORMATION In addition to information that defines the geometry of an object, there is the ability to define relationships between objects or entities of objects. As stated in previous sections, if one object is modified, information is automatically sent to the entities with which there is a relationship, and the entity is appropriately updated.

NONGEOMETRIC INFORMATION Nongeometric information is the attribute information such as density, viscosity, and material type. For example, a specific cavity insert has a specific material type, with a specific vendor catalog number, with a specific cost.

13.1.2 Benefits of Knowledge-Based Engineering

Knowledge-based engineering automates the design process, allowing many more design iterations to be considered. The significant reduction in the design cycle allows for sensitivity analysis and product or process design optimization. Engineering automation can be used effectively in many engineering situations such as automating repetitive designs, centralizing and standardizing design "knowledge," integrating analysis tools, supporting concurrent engineering, and reducing costs (Cinquegrana, 1990a).

Automating Repetitive Designs

Engineers must frequently perform tedious calculations and catalog look-ups. Thus much of the engineer's time is spent reanalyzing and reevaluating their designs. Frequently, new products are just modifications of current products (Cinquegrana, 1990b) designed from similar components. These similar components are analyzed using similar engineering techniques. Knowledge-based engineering automates the engineering techniques for similar entities by encoding them into rules in the product model. This automation is not restricted just to objects but also extends to processes. In a complex design situation, the object may be composed of many entity objects, which may themselves in turn be composed of many entities. A design process, for example, may contain the compilation of many engineering tasks on similar objects.

Centralizing and Standardizing Design "Knowledge"

Knowledge-based systems allow the construction of design "primitives." These design primitives are objects that are routinely used throughout the current product model as well as other product models. The model contains descriptions of the complex design rules, relationships, engineering standards, manufacturing constraints, tooling, process plans, costs, and performance specifications used to design and produce the object. Use of such "primitives" encourages standardization of design components and integration of standard components to establish standard assemblies, which enhances manufacturability.

Integrating Analysis Tools

There are many software analysis packages that assist the engineer in the design process. Each of these analysis packages requires a specific input format to be generated. Depending upon the complexity of the package, this input format may be difficult and/or time-consuming to produce. Knowledge-based systems allow the package-specific input format to be automatically generated from the design specifications.

Concurrent Engineering

In product or process design there are many engineering departments involved. Typically, each engineering department is a separate group that separately and/or sequentially reevaluates the design product or process. In knowledge-based engineering systems, the design parameters and constraints for all engineering departments are incorporated into a single model. When a design alternative is considered in one department, the change is automatically checked against other engineering departments' rules and constraints.

Cost Reduction

Design and manufacturing reports such as bill of materials, CAD drawings, and process plans are typically generated manually. Unnecessary design costs are incurred when engineering changes are made and new reports must be re-generated. Knowledge-based engineering systems automate the generation of all such reports. Short design cycles reduce time to market and reduce product costs.

13.2 KNOWLEDGE-BASED SYSTEMS

13.2.1 Definition of Knowledge-Based Engineering

Knowledge-based engineering is an object-oriented software technology that allows the storage of product or process information as a set of engineering rules, attributes, and requirements. Therefore, the "intelligent" model captures the design intent behind the geometric information. The stored information is called the product model and can be used to generate design outputs such as a bill of materials (BOM), process plans, and numerical control (NC) toolpath information automatically. The "intelligent" model can create a specific design or an instance of that family of parts given the input specification (Cinquegrana, 1990a).

13.2.2 ICAD Design Language

ICAD is a knowledge-based, object-oriented system developed for construction of product design knowledge bases. The system consists of a symbolic language for the description of the design, a graphical interface for display of the geometric behavior of the model, and a relational query system for locating objects and tabular information.

Symbolic Language

The ICAD design language is superset of LISP, a symbolic language which is the preferred language of artificial intelligence research. ICAD incorporates artificial intelligence techniques such as demand-driven evaluation, inheritance symbolic referencing, and nonprocedural declaration.

DEMAND-DRIVEN EVALUATION

A complete design model that covers all functional considerations (rules and constraints) for analysis, CAD, and so on, includes many rules that need not be evaluated to respond to a particular request. The system will not evaluate an attribute if it is not required. For example, during the preliminary design stage, it is not necessary to generate a BOM. When a parameter is evaluated, the system will remember the value and will not be required to recalculate it. This technique is an efficient computing method.

INHERITANCE

The general rule of "inheritance" is: whenever x is a member of a set X, which is a subset of a set Y, any property true of any member Y must be also true for x (Cinquegrana, 1990c). There are two types of inheritance used in artificial intelligence—"part of" inheritance and "kind of" inheritance.

"KIND OF" INHERITANCE "Kind of" inheritance allows the user to define rules for a subclass of objects, and to make these rules available to different classes of objects. To illustrate, the definition of cavity-insert might define the general characteristics such as insert-wall-thickness and shoulder-length. The definition of cavity-insert might be then used to define a more specific cavity insert that has the same general characteristics with additional characteristics. These additional characteristics could be an object-specific characteristic or an "override" of a general characteristic. For example, a round-cavity-insert could be defined which inherits the characteristics of cavity-insert, and creates additional features such as the fact that the insert-radius is half the cavity-insert width.

"PART OF" INHERITANCE "Part of" inheritance allows an object to specify a parameter for its subassemblies and subassemblies of these, and so on. To illustrate, the definition of a runner-layout may define its parts to have a sprue, runner elements, and gate elements. The object model may define the plastic-material-type to be poly(ethylene), which will be inherited by the sprue, runner elements, and gate elements.

SYMBOLIC REFERENCING

The definition of a part can refer to the definition of any other part, using a symbolic referencing chain. This reference chain defines the logical path to any other object. This technique allows the dependencies of one part or another to be defined. For example, the runner-layout pressure-drop definition might read as follows:

```
{13.2} (+  (the :sprue :pressure-drop)
           (the :runner-element :pressure-drop)
           (the :gate-element :pressure-drop)
```

This feature allows the interaction of objects with other related objects and the specification of object features for other objects.

NONPROCEDURAL DECLARATION

In a nonprocedural, or declarative, language, it is not necessary to specify the order in which the rules are executed. The language itself determines the dependence of one rule on another and establishes the program flow. In a procedural language, the program flow must be programmed specifically for the flow dependencies.

13.2.3 ICAD Design Language Structure

All object models are grouped into classes or sets. These classes are made up of subclasses, which are made of sub(subclasses), and so on. In design, it is conventional to break down a complex model into several simpler models. In knowledge-based engineering it facilitates the problem-solving process to break complex model (assembly) hierarchically into subassemblies that consist of standard design entities. Consequently, there are many objects of different types and behaviors that need to be defined and managed. The product "tree structure" represents a method of describing the configuration of mechanical systems and its component entities.

Tree Structure

Complex assemblies are represented by a *tree structure*, also known as the *product structure tree*. Each object in the tree represents a part in the overall assembly and is called a *node*. A *parent node* is a node that has other nodes

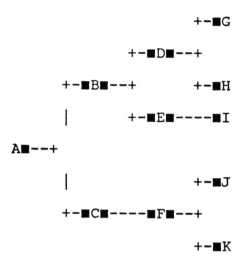

Figure 13.3. Tree structure.

below it in the tree. A node that is directly beneath a node is called a *child* of the parent node. The node that does not have a parent is called the *root* of the tree. Nodes that do not have any children are called *leaves* of the tree. All nodes below a particular node are called *descendants* of that node. Children of the same parent are called *siblings*. Figure 13.3 illustrates a generic tree structure.

In general, the root of the tree, A, represents the overall assembly. The branches of the tree represent subassemblies, and the leaves of the tree represent the physical components of the assembly. Particular branches of the tree such as B and its descendants are referred to as subtrees. The root A has two children, B and C, which are siblings. B is the parent of D and E. C is the parent of F, which has children J and K, which are siblings. Nodes G, H, I, J, and K are leaves of the tree.

It is important to realize that a part is not necessarily a physical object but may be a conceptual object. For example, a set of calculations may be defined as an abstract object that imparts a "kind of" inheritance to a physical object.

DEFINITION OF PRIMITIVE ENTITIES

The design language allows the construction of primitive entities to define both physical and conceptual objects. The main classifications of parameters used to model the behavior of an entity or class of entities are inputs, optional-inputs, defaulted-inputs, attributes, and parts.

INPUTS These parameters define a list of input-attributes that represent the design specifications for the subparts of the assembly. The design specification determines the behavior of the part design; thus different input specifications will create a different entity of the same object class.

OPTIONAL-INPUTS These parameters define a list of features whose default definition should be used if there is no input-specification, that is, inputs with default values. These parameters only need to be specified when the input value is different from the default value.

DEFAULTED-INPUTS These parameters are similar to optional-inputs in that they define input-attributes with default values. They are different in that they will try to inherit a value if no input-specification is given.

ATTRIBUTES These parameters describe an object's physical dimensions, geometric location in space, or other parameters required to generate the model. Some attributes are engineering rules or constraints, while others are input-attributes or constant values for this class of objects.

PARTS These parameters define the subassemblies or subparts of the object or assembly. This establishes the product structure tree of the assembly or object, and the framework for positioning and orienting components relative to one another.

PART–WHOLE DEFAULTING

When an object in the tree requires a value for an attribute that does not appear in its own definition, the object (node) asks its ancestors for the value. This is called part–whole defaulting since certain attributes can be made available as default values to parts in the tree. Part–whole defaulting occurs when an object is defined to be a child of another object or node in the tree.

COORDINATE SYSTEMS

Objects are oriented with respect to a local coordinate system. An object's orientation never changes with respect to the object's local coordinate system. If an object is a subassembly of another object, its orientation and position are viewed with respect to the global coordinate system, or with respect to the root node in the tree. Therefore, the orientation and position of an object depends on the orientation and position of its parent in the tree structure.

Relational Database and Query Facility

The relational database and query facility can be used for both manipulation of tabular information and generation of design reports.

PROCESSING OF TABULAR (CATALOG) DATA

A substantial amount of engineering and design information is in a tabular format. Standard components and raw materials are specified by vendors' catalogs. Company standards and codes are stored in this fashion as well. ICAD contains a facility for constructing these tables and a relational query language for accessing the data.

A query consists of an unlimited sequence of query objects with an associated set of attributes. To illustrate, the D-M-E Catalog contains information on injection mold bases distributed by the Detroit Mold Engineering Company. The query objects described consist of rows of column attribute values such as mold price, a-plate length, and so on. The catalog, once defined, can be used as a query source. Query operators such as SELECT and SORT are filters used to retrieve the appropriate database query object. These query objects are selected to meet all of the criteria specified by the filter. For example, the filter might determine all the mold bases larger than 15 in × 18 in in the query source (which is the D-M-E mold base catalog), sorted in ascending order of mold price. Such a filter might read as follows:

```
{13.3}  (:sort
            (:select D-M-E-mold-base-catalog
            ( > (the-element :mold-size) 1518))
        (( < :price)))
```

GENERATING DESIGN REPORTS

The query source can originate from engineering knowledge bases such as the tree structure of the design. Any part in the tree structure may contain methods for collecting information and formatting a report. For example, the mold base described may include a bill of materials report, which will ask each component of the mold to give details such as catalog number and price. The report specification might also include instructions to perform subtotal and total cost calculations. The report information can be formatted under headings and documented according to the report definition.

Graphical Interface

The "browser" is a graphic tool for examining the behavior of the design knowledge base. It is used to graphically view the behavior of a design entity to a given input specification. The browser is composed of three windows, the 3-D graphics window, the tree display, and the inspector window. These windows retain a command history of the sequential operations performed on a given object, and there is unlimited "undo" capability; that is, any operation or sequence of operations can be revoked at any time to return the object or system to the desired previous state.

GRAPHICS WINDOW In the graphics window, the geometric design created from input-specifications and the rules embodied in the intelligent object can be viewed. The window shows a 3-D wire-frame in standard, zoom, or rotated display at any projection perspective, and is capable of displaying multiple views simultaneously. Multiple graphics windows can be created and subdivided so that alternative designs can be evaluated and compared. The graphics window is continually scaled to accommodate the displayed object(s).

DISPLAY TREE The display tree is a hierarchical tree structure generated from the input-specification. Using a mouse, a user can "navigate" through the three structure and access any object definition. Nodes of the tree can be selectively displayed, and asked to identify design parameters in the object's rule set.

INSPECTOR Each node of the tree structure has an inspector window associated with it. The window displays all of the object's design parameters, or attributes, and their respective values. As discussed previously, the control structure is demand-driven, and consequently does not require all attributes of an object to be evaluated. The unbound values associated with the object can be evaluated and displayed in the inspector window.

13.3 DESIGN EXAMPLE: PLASTIC MOLD DESIGN

It is best to illustrate the usage of a knowledge-based system with an example of mold design. The knowledge base is organized such that the only top level input is the candidate product design to be molded. The knowledge base will prompt the engineer, via the user interface, for any design parameters not yet known to the system. The knowledge base will automate the mold design process through engineering calculations for polymer flow, such as runner balancing and sizing; mechanical elements, such as plate dimensioning; and catalog lookups, for retrieving standard mold components from vendor catalogs.

Once the knowledge base completes the mold design for the candidate product design, the engineer can view the results in the graphical interface. Figure 13.4 displays the result of the mold design for one specific instance. The knowledge base not only represents the design graphically but automatically calculates all the information for each component of the design which the engineer can inspect in the graphical interface. Figure 13.5 illustrates

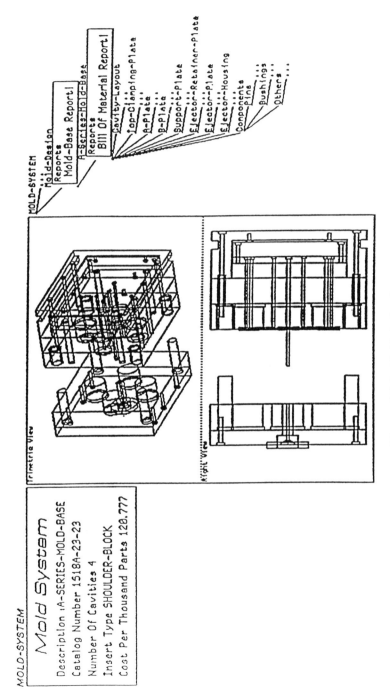

Figure 13.4. Mold design layout.

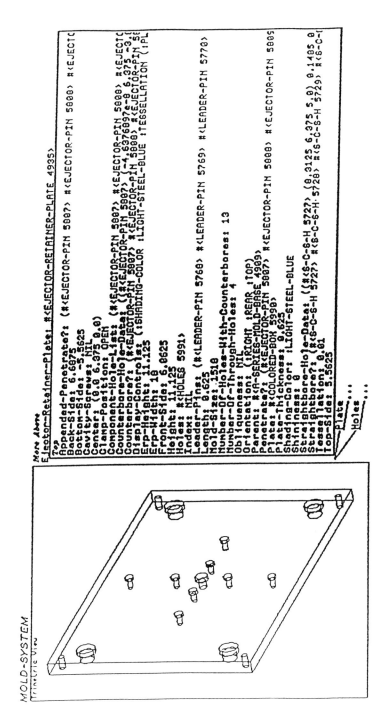

Figure 13.5. Geometric and nongeometric information of mold base.

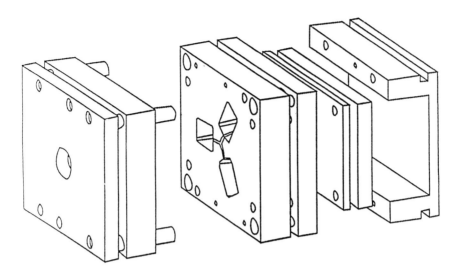

Figure 13.6. Model of mold design assembly.

the geometric (graphical) and nongeometric information of one of the mold components in the mold design.

The output from the knowledge base can be viewed as a wire-frame model or can be displayed as a solid model via the parasolid modeler (Figure 13.6). The knowledge base can easily interface to external (analysis) packages for detailing of the product model. To illustrate interfacing capabilities, the result of the solid model can be sent to a process planning program for tool path generation and generation of the process model for the mold design. Figure 13.7 displays the process plan for the mold base component in Figure 13.5.

13.4 SUMMARY

This chapter describes the differences between the traditional design and engineering process and the knowledge-based engineering process. Though the latter is still in its infancy, it can facilitate the implementation of many of the concepts of concurrent engineering discussed in this book: tolerance analysis, manufacturability requirements, engineering analysis, process planning, cost analysis, and manufacturing documentation.

These inputs, analysis tools and constraints can be combined into the knowledge-based engineering process to provide outputs that are directly

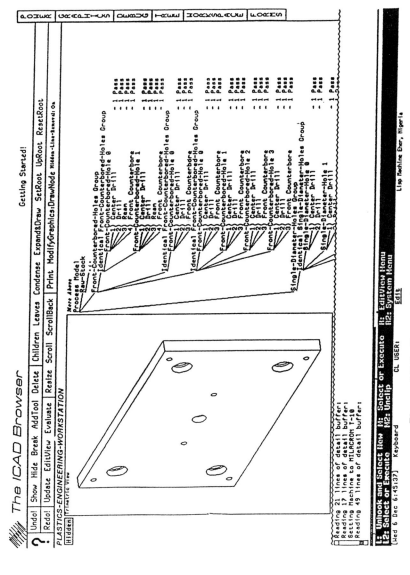

Figure 13.7. Process plan for mold base component.

related to an optimized design based on existing design and manufacturability rules, the goals of the design as outlines by the engineer, and the results of performing engineering analysis on the design.

There are many commercial systems currently available that offer knowledge-based design processes, either on a stand alone basis, or as an adjunct to already established CAD systems. These systems offer many of the features discussed in this book, some specializing in particular manufacturing processes such as plastics or sheet metal or by mathematical modeling and analysis of the design such as thermal flow or finite element analysis. Other systems specialize in the design of a user friendly interface for the engineer. As the knowledge-based design process matures, its future impact will be truly revolutionary in enhancing the concurrent engineering process.

APPENDIX A Complete Model Definition for Bushing Objects

```
;;****************************************************************
;;
;;
;;
;;****************************************************************
;;**************


;;******** HERE IS THE MIXIN DEFINITION ********

(defpart bushing-mixin (base-object)
  :inputs
  (:mold-size)
  :attributes
  (:object-locations (list-elements (the :pins :leader-pins) (the-element))
  :location-list
  (let-streams ((object (in (the :object-locations)))
          (center (the-object object :center))
          (x-coor (get-x center))
          (z-coor (get-z center))
          (y-coor (get-y (the :b-plate :center)))
          (point (make-point x-coor y-coor z-coor))
          (points (collect! point)))
    ((return-when empty? points)))))


;;******** HERE IS THE QUANTIFIED PARENT DEFINITION **********


(defpart shoulder-bushings (bushing-mixin)
  :parts
```

APPENDIX A

```
((shoulder-bushing :type shoulder-bushing
             :quantify (:series 4)
             :mold-size (the :mold-size)
             :center (nth (the-child :index) (the
:location-list)))))


;;******** HERE IS THE BASE COMPONENT DEFINITION
***********************


(defpart shoulder-bushing (billable-part-mixin color-mixin
base-object)
  :inputs
  (:mold-size)
  :query-attributes
  (:shoulder-bushing-query
    (:sort
      (:select
      (:select dme-shoulder-bushing
      (= (the-element :nominal-inner-diameter) (the
:desired-inside-diameter)))
      (>= (the-element :bushing-length) (the :desired-
length)))
      ((< :bushing-length))))
  :attributes
  (:desired-length (the (the :b-plate :length)
   :shading-color *color-20*
   :transparency *transparency-20*
   :shininess *shininess-20*
   :reference 'bushing
   :desired-inside-diameter
   (cond
     ((<=  (the :mold-size) 1008) (inch 0.75))
     ((<n<= (the :mold-size) 1008 1123) (inch 0.875))
     ((<n<= (the :mold-size) 1123 1329) (inch 1))
     ((<n<= (the :mold-size) 1329 1935) (inch 1.25))
     ((<n<= (the :mold-size) 1935 2435) (inch 1.5)))
   :desired-outside-diameter
   (+ (the :desired-inside-diameter)
      (cond ((<= (the :mold-size) 1935) (inch 0.375))
            ((<n<= (the :mold-size) 1935 2435) (inch 0.5))))
   :shoulder-bushing-data (retrieve (the :shoulder-bushing-
query))
   :bushing-length (the :shoulder-bushing-data :bushing-
length
   :inner-diameter (the :shoulder-bushing-data :bushing-
inner-diameter)
   :outer-diameter (the :shoulder-bushing-data :bushing-
outer-diameter)
   :shoulder-diameter (the :shoulder-bushing-data :shoulder-
outer-diameter)
   :shoulder-thickness (the :shoulder-bushing-data
:shoulder-thickness)
   :price (the :shoulder-bushing-data :price)
   :catalog-number (the :shoulder-bushing-data :catalog-
number))
  :parts
  ((shaft   :type cylinder
            :position (:in-front (:from (the :b-plate)
                        (- (- (the :bushing-length)
                              (the :shoulder-thickness)))))
            :radius (half (the :outer-diameter))
            :inner-radius (half (the :inner-diameter))
            :length (- (the :bushing-length)
```

```
                    (the :shoulder-thickness)))
     (head    :type cylinder
         :position (:in-rear (:from (the :shaft) 0))
         :radius (half (the :shoulder-diameter))
         :inner-radius (half (the :inner-diameter))
         :length (the :shoulder-thickness))))


;;******** HERE IS THE CATALOG INFORMATION ON THE SHOULDER
BUSHINGS ************

catalog
{(:defcatalog dme-shoulder-bushing)}
{(:defline
   (:bushing-length
     :catalog-number
     :price
     :nominal-inner-diameter
     :bushing-inner-diameter
     :bushing-outer-diameter
     :shoulder-outer-diameter
     :shoulder-thickness
     ))}
{(:prepare-to-read)}

     .8750   5700      8.10     .7500     .7505    1.1255
 1.3020   .1875
    1.3750   5701      8.80     .7500     .7505    1.1255
 1.3020   .1875
    1.8750   5702      9.50     .7500     .7505    1.1255
 1.3020   .1875
    2.3750   5703     10.20     .7500     .7505    1.1255
 1.3020   .1875
    2.8750   5704     11.00     .7500     .7505    1.1255
 1.3020   .1875
    3.3750   5705     11.70     .7500     .7505    1.1255
 1.3020   .1875
    3.8750   5706     12.40     .7500     .7505    1.1255
 1.3020   .1875
    4.3750   5707     13.10     .7500     .7505    1.1255
 1.3020   .1875
    4.8750   5708     13.80     .7500     .7505    1.1255
 1.3020   .1875
    5.8750   5709     15.20     .7500     .7505    1.1255
 1.3020   .1875
     .8750   5710      8.30     .8750     .8755    1.2505
 1.4270   .1875
    1.3750   5711      9.00     .8750     .8755    1.2505
 1.4270   .1875
    4.8750   5958     38.15    2.5000    2.5005    3.2505
 3.4270   .1875
    5.8750   5960     44.45    2.5000    2.5005    3.2505
 3.4270   .1875
    3.8750   5976     39.40    3.0000    3.0005    3.7505
 3.9900   .5000
    4.8750   5978     47.80    3.0000    3.0005    3.7505
 3.9900   .5000
    5.8750   5980     56.20    3.0000    3.0005    3.7505
 3.9900   .5000
    7.8750   5984     72.90    3.0000    3.0005    3.7505
 3.9900   .5000
```

APPENDIX A 333

```
;;*************** HERE IS THE DEFINITION OF THE ASSEMBLY
SCREWS *************

(defpart screws (base-object)
  :inputs
  (:mold-size
    :blind-pocket-data)
  :query-attributes
  (:cap-screw-catalog
    (:select dme-mold-base-by-size
      (= (the :mold-size) (the-element :size-reference))))
  :attributes
  (:screw-hole-list (the :screw-list)
    :screw-list
      (append
        (the  :blind-pocket-screws :blind-pocket-screw-list)
        (list-elements (the :ejector-plate-screws) (the-element))
        (list-elements (the :ejector-housing-screws) (the-element))
        (list-elements (the :clamp-plate-screws) (the-element)))
     :cap-screw-catalog-data (retrieve (the :cap-screw-catalog))
     :ejector-cap-screw-loc-lat
       (the :cap-screw-catalog-data :ejector-cap-screw-loc-lat)
     :ejector-cap-screw-loc-vert
       (the :cap-screw-catalog-data :ejector-cap-screw-loc-vert)
     :ejector-cap-screw-size
       (the :cap-screw-catalog-data :ejector-cap-screw-size)
     :num-plate-cap-screws
       (the :cap-screw-catalog-data :num-clamp-plate-cap-screws)
     :plate-cap-screw-size
       (the :cap-screw-catalog-data :clamp-plate-cap-screw-size)
     :plate-cap-screw-location-list
       (the :cap-screw-catalog-data :clamp-plate-cap-screw-location-list))
  :parts
  ((blind-pocket-screws :type (if (the :blind-pocket-data)
                                  'blind-pocket-cap-screw
                                  'null-part)
                :blind-pocket-data (the :blind-pocket-data))
   (ejector-plate-screws  :type ejector-plate-cap-screw
                :ejector-cap-screw-loc-lat (the :ejector-cap-screw-loc-lat)
                :ejector-cap-screw-loc-vert (the :ejector-cap-screw-loc-vert)
                :ejector-cap-screw-size (the :ejector-cap-screw-size))
   (ejector-housing-screws :type ejector-housing-cap-screw
                :num-plate-cap-screws (the :num-plate-cap-screws)
                :plate-cap-screw-location-list (the :plate-cap-screw-location-list)
                :plate-cap-screw-size    (the :plate-cap-screw-size))
   (clamp-plate-screws :type clamp-plate-cap-screw
                :num-plate-cap-screws (the :num-plate-cap-screws)
```

```
              :plate-cap-screw-location-list (the :plate-
cap-screw-location-list)
              :plate-cap-screw-size (the :plate-cap-screw-
size))))

  ;;************** HERE IS THE DEFINITION OF A CLASS OF SCREW
  ********************

  (defpart clamp-plate-cap-screw ( color-mixin base-object)
    :inputs
    (:num-plate-cap-screws
      :plate-cap-screw-location-list
      :plate-cap-screw-size)
    :attributes
    (:shininess *shininess-16*
      :transparency *transparency-16*
      :shading-color *color-16*)
    :parts
    ((top-clamp-plate-cap-screw :type cap-screw
                  :quantify (:series (the :num-plate-cap-
screws))
                  :orientation (:rotate :lateral)
                  :center (position-plate-cap-screw
                        (the :num-plate-cap-screws)
```

REFERENCES AND SUGGESTED READING

Belzer, A. and Rosenfeld, L. *Breaking Through the Complexity Barrier*. Cambridge, Mass.: ICAD Publications, 1987.
Cinquegrana, D. *Understanding ICAD System*. Cambridge, Mass.: ICAD Publications, 1990a.
Cinquegrana, D. *Knowledge-Based Injection Mold Design Automation*, D.Eng. Thesis, University of Lowell, Lowell, Mass., 1990b.
Cinquegrana, D. "Intelligent CAD automated plastics injection mold design." *Mechanical Engineering*, July 1990c.
Cinquegrana, D. "Mold optimization using rules-based software," Antec, 1990d.
Manzione, L. *Applications of Computer Aided Engineering in Injection Molding*. New York: Hanser, 1987.
Potter, C. "Smart drafting systems." *Computer-Aided Engineering*, Feb. 1989.
Robinson, P. "Design debate." *Computer Graphics World*, April 1989.

Index

Absence weight 157
AFD process 156
  opposition points to 184–5
Affinity diagram 155
Analysis of variance (ANOVA) 127, 141–3
Angularity specification 219
ANSI (American National Standard Institute) Y14.5M-1982 208
Answering machines 288
Apple Computers 33, 307
Assemblability evaluation method (AEM) 240
Assembly
  base and stack 51–2
  flexible 193
  manual process 192, 194–6
  minimal directions of 52
  minimum levels of 52
  process 192–4
  quality of process 78
Asymmetrical parts 58–60
ATE (Automatic test equipment) 280
Attitude tolerances 218
Attractive quality 150
Attribute charts 79, 121
Attributes 323
Auto industry 4
Automatic ground vehicles 190
Automatic guided vehicles 197
Automatic Parts Orienting System (APOS) 201–2
Automation 193
  costs 196
  design for 196–203, 318
  design parts for 198
  flexible 193, 203
  integration of 199
  level of 197
  plan 196, 197

successful implementation of 196
  system design 198
Axiomatic corollaries 50
Axiomatic theory of design 49–50
Axioms of design 49

Barcodes 280
Base and stack assembly 51–2
Bathtub curve 259
Bilateral tolerance 222
Bill of materials (BOM) 319, 320
Bonding process, optimization of 133–5
Boolean logic functionality 258
Boothroyd/Dewhurst method 195
Brainstorming 88–90, 125, 155
Breadboard 39
Browser graphic tool 324
Bug chart 45
Burn-in 260–2
Bushing objects, model definition for 330–4
Business plan 34

C charts 79
Cables 56
CAD 35, 108, 191, 238, 258
  and tolerance analysis 228
  limitations of traditional systems 315–19
CAD/CAM systems 315
CAE. *See* Computer-aided engineering (*CAE*)
CAE/CAD equipment 110
CAE/CAD/CAM 57
CAE/CAD/CIM 18
CAE-to-manufacturing process analysis 300
Calibration of production equipment 121
Capital equipment purchases 44
Cathode ray tubes (*CRTs*) 50
Cause and effect diagrams 90
Checksheets 90–2

335

CIM 297, 311
 definition 311–12
 goals 312
Classical design of experiments 144
Communications 288
Competition 7–8, 157
Competitive position 190
Compliance 61
Composite tolerance 219
Computer-aided design. *See* CAD
Computer-aided engineering (CAE) 22, 40, 107, 108
 see also CAE
Computer analysis packages 258
Computer-integrated manufacturing. *See* CIM
Computer simulation 144, 198, 258
Concurrent engineering 2, 319
 as competitive weapon 7–8
 benefits of 104
 definition 1
 functional roles in 104–7
 implementing 3, 104
 long-term benefits of 103
 measuring 115
 need for 4–6
 new products 18–22
 organizing for 109–15
 ownership of 104
 PCB design 189
 results 1, 4, 7, 19–22
 role of 3
  design engineering department 104–5
  manufacturing engineering department 105–6
  other departments 106–7
 techniques of 3
 tools 3, 7
Confounding effect 138
Construction industry 36
Contamination effects 273
Context diagram 9
Continuous process improvements (CPI) 2, 122
Contractual agreements 263
Control charts 79, 121, 199
 calculations 84–5
 factors 83
 flow diagram 93
 generation of 82–5
 interpretation of 86
 limits 83, 85
 shortcomings in using 81
Cooperation 3
Coordinate systems 323
Cost-reduction programs 2
Costs
 automation 196
 information technology 284, 304–5
 knowledge-based systems 319
 life cycle 6, 262
 printed circuit boards 244–6
 robotics 196
Counterpart characteristics 152
Coupled designs, decoupling of 50–1
Creativity 110
Cumulative distribution function 72
Customer demands 155, 156
Customer-driven engineering 147–87
Customer expectations 150, 262
Customer importance rating 156
Customer loyalty 5
Customer needs 5, 150
Customer requirements 5, 277
Customer satisfaction, levels of 149
Customer verbatims 155
Cylindrical tolerance zones 211–13

Data analysis 127, 133–5
 parameter interactions with 139
Data collection 289
Data dictionary 10
Data flow diagrams 9–18
Data presentation 289
Data processing 310
 tabular (catalog) 324
Databases 57
Datums
 and datum features 213–16
 cylindrical features as 216
 partial 216
 permanent 216
 temporary 216
Decision-tree analysis (DTA) 266, 268
Defaulted-inputs 323
Defect analysis, printed circuit boards (PCB) 248–9
Defect data location checksheet 92

INDEX

Defect data recording checksheet 91
Defect generation 122
Defect rates 74–9, 99–100
Defect tracking from field failures 273–4
Defects per unit (DPU) 76–7
Demand-driven evaluation 320
Demanded weight 157
Design
  and reliability 263–4
  automation 196–203, 318
  axiomatic theory of 49–50
  axioms of 49
  efficiency 65, 195, 204
  function 49
  goals 65–6
  guidelines 50–62, 107–9
    in-house versus supplier capability 108
  knowledge-based systems 318, 325–8
  matrix 49
  metrics 115
  plastic mold 325–8
  product 123, 221
  production 221
  requirements 152
  review process 265–6
  traditional 316
  validation 258
  versus life cycle costs 262
  see also Robust design; under Printed circuit boards
Design cycle 6
Design engineering department 104–5
Design engineers 109–10
Design for assembly (DFA) 48
Design for manufacture (DFM) 2, 36
  and design complexity 235
  assembly analysis 205
  Boothroyd and Dewhurst system 240
  concepts of 1–23
  cost metric 244–6
  cost model 245
  critical function definition 299–303
  efficiency 195, 204–5
  enabling support technologies 294
  evolution 279, 288, 294
  guidelines 48, 188, 279
  history 240
  implementation 255–6, 279, 294, 307, 307–11
  implementation team 298–9
  information flows 280–3
  information technology in. *See* Information technology
  interactions and trade-offs between design elements 241
  lessons learned 311–12
  management approval 305–7
  management commitment 287–8
  management role 297–8
  metrics definition 299
  new products 240
  pilot project 307
  planning implementation of support technology 296–307
  principles of 48
  printed circuit boards (PCBs) 235–57
    complexity metric 246–53
    design rules 256
    effect of feature selection yield 246
    evolution of 240–1
    overall process 254–5
    production yields 250
    program development 256
    support plan 257
    yield loss 248
    yield prediction model 247, 253
  program components 276
  program requirements 242–53
  quantitative 240
  relative cost of alternative designs 244–6
  systematic approach 240
  tools for 276–314
  ultimate goal 257
Design of experiments (DOE) 26, 122
Design project, phases 39–41
Design ratings 188
  for manual assembly 194–6
Design systems and QFD 150–1
Development cycles 5
Development plan 34
Discounted cash flow method 42, 304
Display tree 325
Dual-in-line package (DIP) 237

Electrical engineering CAD (EECAD) systems 293
Electrical failures 272
Electrical noise
  optimizing effect of 139–41

Electrical noise *continued*
  reduction of 144
Electronic mail (E-mail) 288
Electronic products, manufacturing process for 189
Electronics industry 27
Electrostatic discharge (ESD) protection 272
E-mail 288
Engineering model 40
Engineering support equipment 44
Environmental related failures 273
Equipment parameters 121
Equivalent IC (EIC) density 254
Errors 61
Exciting quality 150
Exclusive or (XOR) relationship 136–7

Failure analysis 280
Failure mode and effect analysis (FMEA) 266
Failure rate 259–60, 262
  versus time 261
Fastening design 55
Fault tree analysis (FTA) 266
Feature control frame 219
Feedback 278, 279
Field failures, defect tracking from 273–4
Field service department 107
Field tests 271
Final assembly adjustments 50
Final product assembly and test times 66
Financial analysis 35
First-time yield (FTY) 76–9
Fishbone diagram 90
Flexible assembly 193
Flexible automation 193, 203
Flexible items 61–2
Flexible manufacturing systems (FMS) 190
Flowcharts 92–3
Form controls 218
Four Ms 90
Functional requirements 51–2
Functional testing 258
Functionality concept 221

Gantt charts 44
General Electric 195

Geometric dimensioning and tolerance analysis (GDT) 208–34
  basic elements of 209–10
  concepts and symbols 210
  controls 217–19
  definition 209–10
  element categories 210
  see also Tolerance analysis
Geometric information 317, 328
Goals, setting and measuring 65–6
Graphical interface 324–5
Graphics window 325
Group technology 57
  information 292

Hewlett-Packard 7–8, 33, 196
Hidden features 60
Histograms 95–6
Hitachi 195

IBM Proprinter 48, 50, 62–5, 195, 205
ICAD design language 319–25
Improvement plan 157
Improvement ratio 157
Incoming component quality 78
Independence Axiom 49
"Infant mortality" period 259, 260
Information Axiom 49
Information flows 300
  and DFM 280–3
  engineering 282
Information technology
  changing role of 277–80
  configuration 310
  costs 284, 304–5
  guidelines for selection of 307
  implementation to support DFM 296–307
  infrastructure of 287
  integration 309–10
  procurement process 309
  requirements for DFM 283–96
  role in DFM 276–314
  support 310–11
  training 309
Information transfers 278, 299
Inheritance
  general rule of 320–1

INDEX   339

Inheritance *continued*
   kind of   320
   part of   320
Inputs   323
Inspection stations   199
Inspector window   325
Institute of Printed Circuits (IPC)   107, 189, 239
Integrated circuits (ICs)   76
   designs   40–1
   failure rates   260
Investment justification guidelines   285
Ishikawa diagram   90, 300
Iterative process   33–4, 36

Japan   4
Just-in-time (JIT)   2, 26, 260

Kano model   149
KISS (Keep it Simple, Stupid) principle   49
Knowledge-based systems   280–3, 293, 316
   associative geometry   317
   benefits of   318
   cost reduction   319
   definition   319
   design example   325–8
   design primitives   318
   integrating analysis tools   318

Labels   55–6
Laser thermal curing   191
Learning curve   22
Learning process   278, 279
Least material condition (LMC)   216
Levelling process   9
Leveraged effect of design cycle on life cycle costs   6
Liability and product performance   263
Life cycle   5
   costs   6, 262
   model   25–7
   product operating profit   44
   profit sensitivity over   38
   total   263
Life tests   271
Light emitting diodes (LEDs)   51
Limit dimensioning   222–3

Line efficiency   199
LISP   320
Locating features   211
Location checksheet   92
Location control   219
Location tolerancing   211
Loss function   123, 128–31

Manual assembly process
   adaptable   192
   design ratings for   194–6
Manufacturing engineering department   105–6
Manufacturing plan   35
Manufacturing process
   and reliability   271–3
   and tolerance analysis   228
   continuous improvements   86–97
   for electronic products   189
   quality level   78
   variability   70–9, 269
   measurement and control   79–97
Manufacturing requirement planning   287
Margin tests   271
Market analysis   34
Market development   25
Market-in design approach   185
Market pressures   277
Marketing department   106
Marketing strategy   34
Material Requirements Planning   315
Materials properties   121
Matrix management   39
Matrix of matrices   153
Matrix relationships   154
Maximum material condition (MMC)   216
Mean logistic delay time (MLDT)   285
Mean time between failures. *See* MTBF
Mean time to repair. *See* MTTR
Mechanical failures   272
Mechanical review   268
Mercury Computers   8
Methods time measurement (MTM)   194, 196
Mini-spec   11
MLDT (mean logistic delay time)   285
Modular design, subsystem   51
Modular parts   57
Monitoring   199

Motherboard 189
MTBF (mean time between failures) 53, 66, 262–5, 285
  component-based 265
MTM (methods time measurement) 194, 196
MTTR (mean time to repair) 262–4, 285
Multifunctional parts 57
Multistep manufacturing process 76–7

National Bureau of Standards (NBS) 121
National Electronic Packaging and Production Conference (NEPCON) 190
New component qualification process 269–71
New product introduction (NPI) team 280
New products
  concurrent engineering 18–22
  design and development process 24–47
  design for manufacture (DFM) 240
  development 2, 6
  introducing 3
  reliability estimates for 271
  reliability objectives of 262–3
  requirements for 5
Nodes 321–2
Nongeometric information 317–19, 328
Nonprocedural declaration 321
NP charts 79
Number of parts 52–5, 64
Number of suppliers 66
Numerical control (NC) 300–3

Object model, definition of 317
Obsolescence of electronic components 30
Off-line control 121
On-line control 121
Optional-inputs 323
Organizing for concurrent engineering 109–15
Orientation controls 218
Orthogonal arrays 126, 131–5
  parameter interaction in 136–8
Osborne I computer 32
Osborne II computer 32

P charts 79
P-spec 11
Packaging 199
Palletizing efficiency 201, 203
Parameter interactions
  in orthogonal arrays 136–8
  with data analysis 139
Parameter selection 124
Pareto analysis 262
Pareto charts 94, 199, 289, 300
Part design analysis checklist 53–4
Part numbers 57, 66
Part–whole defaulting 323
Parts 323
Parts per million (PPM) 76–7
People metrics 115
Performance evaluation 3
Performance measures 253–4
Pert charts 44
Plastic mold design, design example 325–8
Plastic parts 191–2
Polaroid Corporation 201
Potential failure modes 266
Potential problem analysis 266
Pre-production prototype 40
Price performance curves 27
Primitive entities 322–3
Primitive process 11
Printed circuit boards (PCB) 8, 22, 40–1, 51, 71, 76, 77, 129–31, 189–200
  connectivity performance model 254
  cost metric 244–6
  cost model 245
  defect analysis 248–9
  design alternatives case study 238–9
  design for manufacture (DFM) 235–57
  design interactions 239
  design selection criteria 239
  design stages 237–41
  electrical performance 254
  failure rates 260
  performance measures 253–4
  see also Design for manufacture (DFM)
Probability distributions 74–9
Problem definition 124
Problem-solving techniques 300
Process capability, and statistical analysis 227–8
Process capability index 61, 65, 68–71, 85, 98–100

INDEX

Process design 123
Process planning 315
Process specification 11
Producibility analysis, rule-based 292–3
Product champion 33
Product data sheet 43
Product definition 151
Product design process 123, 221
Product development
  activities and areas of responsibility 37
  commodity stage 26
  concept stage 33
  flow diagram 33
  growth stage 25–6
  managing 36–9
  maturity period 26
  role of technology in 27–32
  slippage effects on life cycle sales 32
  startup stage 25
  team building 110–12
  team conflict 113–14
  team development stages 112–13
  team leader 114–15
  total process 32–9
Product life cycle. *See* Life cycle
Product-out design mentality 185
Product qualification 271
Product reliability systems 259–64
Product specifications 5
Product strategy 33
Product structure tree 321–2
Product support plan 35
Product testing for reliability enhancement 269–73
Production design 221
Production phase metrics 115
Production variability, reduction of 120
Profile controls 218
Profit sensitivity over product life cycle 38
Project management techniques 44
Project team 124
Project tracking and control 42–5
Project update tools 45
Prototype or mock-up 35
  design 221

QFD. *See* Quality function deployment
Quad flat pack (QFP) 237

Quality analysis, statistical. *See* Statistical quality analysis
Quality characteristics 152
Quality department 106
Quality engineering, on-line and off-line 120–2
Quality function deployment (QFD) 5, 39, 49, 106, 147–87
  and design systems 150–1
  applications 148
  benefits of 185
  breakthrough phase 153
  case study 156–61
  companies using 148
  definition 148–50
  descriptive phase 152
  facilitators 155
  glossary of terms 185–6
  guidelines 184
  history 147–8
  implementation 154, 184
  key to success 149
  long-term approach 185
  organization phase 151–2
  phases of 151–5
  selections 154–5
Quality improvement 131
Quality level 66

R charts 79, 82, 84, 121
Radiofrequency interference 144
Recording checksheet 90–2
Reflow process 125
Regardless of feature size (RFS) 217
Reject level 71–4, 122
Relational database and query facility 323
Reliability
  and design 263–4
  and manufacturing process 271–3
  design review for 265–6
  enhancement 258–75
    design tools and techniques for 264–9
    product testing for 269–73
  estimates for new products 271
  objectives of new products 262–3
  product systems 259–64
  stress analysis for 267–9
Report generation 324

Research and development (R & D) department 2
Return factor (RF) 42, 115
Return on assets (aROA) 304
Return on investment (ROI) 35–6, 304
Risk priority number (RPN) 266
Robotics 190, 193–4, 196–203
  costs 196
Robust design 120–46
  comparison with classical design of experiments 144
  conducting an experiment 124–8
  experiment design 126–7
  in black magic problems 144
  in engineering design projects 143–4
  prediction and confirmation of experiments 128
  statistical analysis of experiments 141–3
  techniques 122–8
  tool set 128–43
Rule-based producibility analysis 292–3
Run charts 96–7
Runout controls 219

Sales points, characterization of 157
Scatter diagrams 95
Screening experiments 125
Self-diagnosis capability 50–1
Self-locating features 60–1
Sheet metal fabrication 190–1
Shock effects 273
Signal-to-noise ratio 123, 135
Society for Manufacturing Engineers (SME) 107
Software analysis packages 258
Software components 40, 41
Software enhancements 41
Soldering quality 78–9
Sony Multi-Assembly Robot Technology (SMART) robotic assembly line 201
Specifications 269
  limits 65
SPICE modeling 258
Spider chart 45
Square tolerance zone 211
Standard normal distribution (SND) 72
Standard parts 65
Standardization 56–7
Startup 25

Statistical analysis 280
Statistical process control (SPC) 5, 61, 68, 86, 227, 289
Statistical quality analysis (SQA) 289, 291–3, 300
Statistical quality control (SQC) 289
Statistical tolerance analysis 224–8
Stereo lithography 191
Stress analysis for reliability 267–9
Stress testing 269
Structure charts 3, 8–18
Structured analysis 3, 8–9
  case study 13–18
Structured designs 8–9
Subsystem modular design 51
Surface flatness 218
Surface-mounted technology (SMT) 105
Symbolic language 320
Symbolic referencing 321
Symmetrical parts 57
System design, automation 198
Systems engineering (top-down) approach 299
Systems integration 310

Tabular (catalog) data processing 324
Taguchi process statistics 136
Tape automatic bonding (TAB) 105
Team and teamwork 3, 110, 110–15, 124, 280, 298–9
Thermal shock effects 273
Time-and-motion study 194
Time series graphs 96–7
Time to market (TTM) 103, 235, 238, 276, 296
Tolerance 60–2
  bilateral 222
  composite 219
  data 280–1
  definition 222
  limits 68–70, 72
  of form 223
  of size 222
  plus or minus 222
  stackup 222, 268
  unilateral 222
  zones 218
Tolerance analysis 220–33
  and CAD 228

INDEX

Tolerance analysis *continued*
  and manufacturing processes 228
  best-case 223
  case study 228–33
  definition 222
  extreme-case 223, 224–6
  need for 221–3
  statistical 224–8
  worst-case 223–6
  see also Geometric dimensioning and tolerance analysis (GDT)
Top down partitioning 9
Total life cycle 263
Total quality control (TQC) 262
Total quality management (TQM) 2, 5
Training 309
Transportability, design for 198
Tree diagram 155
Tree structure 321–2
True positioning 211
Turn-on level 77–9

U charts 79
Unilateral tolerancing 222
Universal Instrument Corporation 107
"Useful life" period 259

Validation
  design 258
  testing 271
Value engineering 2
Variability
  manufacturing process 70–97, 269
  production 120
  reduction 120–46
Variable charts 79
Verified design 41
Vibration effects 273
Vibratory feed stations 199, 201
Voice mail 288

Warranty costs 66, 264
Warranty data 274
"Wearout" phase 259
Work measurement 194
Workmanship defects 271
Worst-case study 144

X-bar charts 79, 82, 84, 121

Z-based distribution 74